ADVANCES IN MULTI-PHOTON
PROCESSES AND SPECTROSCOPY VOL. 2

ADVANCES IN MULTI-PHOTON PROCESSES AND SPECTROSCOPY

Edited by S H Lin

Volume I

Contents: Atomic Multiphoton Processes by *J. H. Eberly and J. Krasinski;* Some Studies on Laser Multiphoton Ionization and Multiphoton Ionization Dissociation of Polyatomic Molecules by *D. A. Gobeli, J. J. Yang and M. A. El-Sayed;* Laser-Induced Molecular Dynamics: Rate Processes in the Gas Phase and at Solid Surfaces by *J. T. Lin, M. Hutchinson and T. F. George;* Multiphoton Processes by Visible and UV Lasers by *I. Tanaka and M. Kawasaki;* Applications of Raman Spectroscopy to Structural and Conformational Problems by *J. Laane;* Theory of Laser-Stimulated Surface Processes: Master Equation Approach by *B. Fain, A. R. Ziv, G. S. Wu and S. H. Lin.*

ADVANCES IN MULTI-PHOTON PROCESSES AND SPECTROSCOPY

Volume 2

Edited by

S H Lin
Department of Chemistry
Arizona State University
Temple, Arizona 85287, USA

World Scientific

Published by

World Scientific Publishing Co Pte Ltd.
P O Box 128, Farrer Road, Singapore 9128

Library of Congress Cataloging-in-Publication data is available.

ADVANCES IN MULTI-PHOTON PROCESSES AND SPECTROSCOPY VOL. 2

ISBN 9971-50-085-X

Printed in Singapore by Kim Hup Lee Printing Co Pte Ltd.

PREFACE

In view of the recent rapid growth in both experimental and theoretical studies of multiphoton processes and multiphoton spectroscopy of atoms, ions and molecules in chemistry, physics, biology, materials sciences, etc., it is desirable to publish an Advanced Series that contains review papers readable not only by active researchers in these areas but also by those who are not experts in the field but intend to enter the field. The present series attempts to serve this purpose. In this series, both theory and experiment are equally emphasized, and each review article is written in a self-contained manner by the experts in the area so that the readers can grasp the knowledge in the area without too much preparation.

The topics covered in this volume are theory of multiphoton processes in molecules (Chapter 1), photochemistry, and photophysics of molecular IR multiphoton excitation (Chapter 2), theoretical treatment of intense field multiphoton processes (Chapter 3), time-resolved resonance Raman spectroscopy (Chapter 4) and application of resonance enhanced multiphoton ionization to radicals (Chapter 5). The editor wishes to thank the authors for their important contributions. It is hoped that the collection of topics in this volume will prove to be useful, valuable and stimulating not only to active researchers but also to other scientists in the areas of biology, chemistry, materials science and physics.

S. H. Lin

CONTENTS

1. THEORY OF MOLECULAR MULTIPHOTON TRANSITIONS

Y. Fujimura

Department of Chemistry, Faculty of Science
Tohoku University, Sendai 980, JAPAN

2

Contents

I. Introduction

In recent years, there has been a great advance in the visible and/or UV multiphoton spectroscopy to investigate mechanisms of higher order molecule-photon interactions, structure of vibronic states, and the dynamic behaviors taking place in the electronically excited states of molecules.[1-8]

In this chapter, we are concerned with theoretical treatments of resonant molecular multiphoton transitions with emphasis on our work carried out during the past five years.

A multiphoton process is said to be in resonance if the energy of one or several quanta of photon is close to a transition energy between relevant two electronic states. In the resonant multiphoton transition, effects of dampings in the resonant states make a significant contribution to the multiphoton process. These effects originate from intramolecular interactions such as nonadiabatic, vibronic, rovibronic interactions, and/or from intermolecular interactions including inelastic and elastic interactions. One of the simplest method for derivation of an expression for the multiphoton transition probability in the resonant case is to use the expression derived in the non-resonant case by introducing the damping constants into the denominator of the cross section. The transition probability is expressed in the perturbation theory as[9]

$$W^{(n)} = \frac{\sigma^{(n)} I^n}{(\hbar\omega_R)^n} \qquad (1.1)$$

where $\sigma^{(n)}$, I and ω_R denote the cross section for n-photon processes in units of $cm^{2n}s^{n-1}$, the laser intensity in units of $erg\ cm^{-2}s^{-1}$, and the laser frequency, respectively. Equation (1.1) expresses simultaneous n-photon processes. Each energy denominator involved in $\sigma^{(n)}$ takes a real form which is expressed in terms of the laser frequency, and the frequency difference between relevant two levels. By using the treatment mentioned above, we can avoid divergence of the transition probability and calculate the magnitude of the transition probability for the simultaneous resonant multiphoton process. It is, however, insufficient to discuss the mechanism of resonant multiphoton transitions. In that treatment, for example, sequential multiphoton transitions can not be considered. In order to discuss the mechanism of the resonant multiphoton processes of molecules, it is necessary to properly take into account the damping effects. These effects have to be considered prior to take time $t \to \infty$ in deriving the transition probability expression in the perturbation theory.

In Sec. II, a theory of the resonant multiphoton transitions, based on the master equation approach is first presented to take into account the damping effects. An expression for the dephasing constant is derived. It is interesting to investigate the resonant multiphoton transition and the resonant light scattering of molecules via coherently excited resonant states. The creation of the coherent states has been proved by observing quantum beats in fluorescence of molecules. A theoretical treatment about the time dependence of vibronic states after the coherent excitation is given in the density matrix formalism in Sec. II. Transition rates of two-photon transitions via the coherent excited resonant states are finally formulated in Sec. II.

In Sec. III, several applications of the theoretical treatment described in Sec. II are presented: Mechanisms of resonant two- and three-photon transitions, quantum beats in fluorescence, and time-resolved resonance Raman scattering from molecules with a nonequilibrium vibronic distribution are discussed.

II. General theory

II-1. Derivation of master equation

In this section, we derive the master equation which describes resonant multiphoton processes of molecules interacting with the heat bath. The master equation can be obtained by using the projection operator method,[10-12] or the cumulant expansion method.[13-16] We adopt the projection operator method here.

We consider a total system consisting of molecules (denoted by S), the heat bath (B) and the radiation field (R). The wave functions of the total system $\psi_i(t)$ satisfy the time-dependent Schrödinger equation:

$$i\hbar \frac{\partial}{\partial t}|\psi_i(t)> = \hat{H}|\psi_i(t)> . \qquad (2.1)$$

Rather than describing the time-dependent behavior of the system in terms of the wave functions, it is convenient to use the density matrix defined by

$$\rho(t) = \sum_i N_i |\psi_i(t)><\psi_i(t)| , \qquad (2.2)$$

where N_i represent the weighting factor. The density matrix can be expanded in terms of any basis set $\{|m><n|\}$ in the Liouville space (double space) as

$$\rho(t) = \sum_m \sum_n \rho_{mn}(t) |m><n| , \qquad (2.3)$$

and the vector in the Liouville space is defined as

$$|\rho(t)>> = \sum_m \sum_n \rho_{mn}(t) |mn>> . \qquad (2.3a)$$

The expansion coefficients $\rho_{mn}(t)$ which the projection of $\rho(t)$ on the basis vector $|mn>>$ in the Liouville space, are given by

$$\rho_{mn}(t) = \sum_i N_i C_{im}(t) C_{in}^*(t) , \qquad (2.4)$$

which is obtained by applying the following expansion of the ordinary wave vector to Eq. (2.2)

$$|\psi_i(t)> = \sum_m C_{im}(t)|m> . \qquad (2.5)$$

The time evolution of the total system is determined by the Liouville equation for the density matrix $\rho(t)$ of the system:

$$-\frac{\partial}{\partial t}\rho(t) = i\hat{L}\rho(t) , \qquad (2.6)$$

where the Liouville operator $\hat{L}$ is defined as

$$\hat{L} = [\hat{H}, \ ]/\hbar , \qquad (2.7)$$

which can be derived by using Eqs. (2.1) and (2.2), and the Hamiltonian of the total system is given by

$$\hat{H} = \hat{H}_0 + \hat{H}_{SR}' + \hat{H}_{SB}' , \qquad (2.8)$$

with $\quad \hat{H}_0 = \hat{H}_S + \hat{H}_B + \hat{H}_R . \qquad (2.9)$

Here, $\hat{H}_S$, $\hat{H}_B$, and $\hat{H}_R$ denote the Hamiltonians of the molecules, heat bath and photon field, respectively. $\hat{H}'_{SR}$ and $\hat{H}'_{SB}$ represent the interaction Hamiltonian between the molecule and photon field, and that between the molecule and heat bath, respectively. We have assumed that there is no interaction between the heat bath and photon field.

Applying the Laplace transformation

$$\rho(p) = \int_0^\infty dt \; \rho(t) e^{-pt} , \qquad (2.10)$$

to Eq. (2.6), we obtain

$$(p+i\hat{L})\rho(p) = \rho(0) , \qquad (2.11)$$

where $\rho(0)$ is the density matrix $\rho(t)$ at $t = 0$, and is assumed to be given by $\rho(0) = \rho^{(S)}(0)\rho^{(B)}(0)\rho^{(R)}(0)$, that is, the subsystems of the total system are independent each other at $t = 0$. It is assumed that these density matrices at $t = 0$ have zero off-diagonal matrix elements.

We are now in a position to find the equation of motion for the reduced density matrix of the molecules. This can be accomplished by applying two types of the projection operators to the density matrix equation of the total system. These are the projection operators which trace out the heat bath variables and photon field ones. In order to tracing out the heat bath variables, we define the projection operator

$$\hat{A} = \rho^{(B)}(0)\text{Tr}_B \ , \qquad\qquad (2.12)$$

in which $\hat{A}^2 = \hat{A}$. Applying $\hat{A}$ and $1-\hat{A}$ to Eq. (2.11) yields

$$\hat{A}[p+i\hat{L}]\hat{A}\rho(p) + \hat{A}[p+i\hat{L}](1-\hat{A})\hat{\rho}(p) = \hat{A}\rho(0) \ , \qquad (2.13)$$

and

$$(1-\hat{A})[p+i\hat{L}]\hat{A}\rho(p) + (1-\hat{A})[p+i\hat{L}](1-\hat{A})\rho(p) = 0 \ , \qquad (2.14)$$

respectively. Substituting $(1-\hat{A})\hat{\rho}(p)$ of Eq. (2.14) into Eq. (2.13), we obtain

$$A[p+i\hat{L}_0+i\hat{L}_{SR}'+\textstyle\sum(p)]\hat{A}\hat{\rho}(p) = \hat{A}\rho(0) \ , \qquad (2.15)$$

where

$$\textstyle\sum(p) = i\hat{L}_{SB}' + \hat{L}_{SB}'(1-\hat{A})[p+i\hat{L}]^{-1}(1-\hat{A})\hat{L}_{SB}' \ , \qquad (2.16)$$

or

$$\textstyle\sum(p) = i\hat{L}_{SB}' - i\hat{L}_{SB}'(1-A)[p+i\hat{L}^0+i\hat{L}_{SR}']^{-1}(1-\hat{A})\textstyle\sum(p) \ , \qquad (2.17)$$

and $\hat{L}^0 = [\hat{H}_0, \ ]/\hbar$ and $\hat{L}_{SB}' = [\hat{H}_{SB}', \ ]/\hbar$. Equation (2.15) can be rewritten as

$$[p+i\hat{L}^0_{SR}+i\hat{L}'_{SR}+<\sum(p)>_B]\rho^{(SR)}(p) = \rho^{(SR)}(0) , \qquad (2.18)$$

where

$$\hat{L}^0_{SR} = [\hat{H}^0_S+\hat{H}^0_R, \quad]/\hbar ,$$

$$\rho^{(SR)}(p) = Tr_B\rho(p) , \qquad (2.19)$$

and

$$<\sum(p)>_B = Tr_B\sum(p)\rho^{(B)}(0) . \qquad (2.20)$$

Neglecting the contribution of the molecule-photon field interaction in the self-energy operator of Eq. (2.17) and within the second Born approximation to the molecule-heat bath interaction, we can express Eq. (2.20) as

$$<\sum(p)>_B \simeq Tr_B\hat{L}'_{SB}[p+i\hat{L}^0]^{-1}\hat{L}_{SB}\rho^{(B)}(0) . \qquad (2.21)$$

We next trace out the radiation field variables by applying the projection operator defined by

$$\alpha = \rho^{(R)}(0)Tr_R , \qquad (2.22)$$

to Eq. (2.18) with Eq. (2.21). By using the same procedure described above, we obtain

$$[p+i\hat{L}_S+<<\textstyle\sum(p)>_B>_R+<\pi(p)>_R]\sigma(p) = \sigma(0) , \qquad (2.23)$$

where

$$<\cdots>_R = Tr_R\{\cdots\rho^{(R)}(0)\} ,$$

$$\sigma(p) = Tr_R\rho^{(SR)}(p) ,$$

and

$$\pi(p) = i\hat{L}'_{SR} + \hat{L}'_{SR}(1-\alpha)G_{SR}(p)(1-\alpha)\hat{L}'_{SR} , \qquad (2.24)$$

or

$$\pi(p) = i\hat{L}'_{SR} - i\hat{L}'_{SR}(1-\alpha)G_{SR}(p)(1-\alpha)\pi(p) , \qquad (2.25)$$

with

$$G_{SR}(p) = [p+i\hat{L}^0_{SR}+<\textstyle\sum(p)>_B]^{-1} . \qquad (2.26)$$

Using Eq. (2.25), we expand Eq. (2.23) as

$$[p+i\hat{L}_S+<\pi^{(0)}(p)>_R+<\pi^{(1)}(p)>_R+<\pi^{(2)}(p)>_R+\cdots]\sigma(p) = \sigma(0) ,$$
$$\qquad (2.27)$$

where $<\pi^{(n)}(p)>_R$ represent the transition operators for the n-photon processes in the P representation, and are given as

$$\pi^{(0)}(p) = <\hat{\sum}(p)>_B \, , \qquad (2.28)$$

$$\pi^{(1)}(p) = \hat{L}'_{SR}G_{SR}(p)\hat{L}'_{SR} \, , \qquad (2.29)$$

$$\pi^{(2)}(p) = -\hat{L}'_{SR}G_{SR}(p)\hat{L}'_{SR}G_{SR}(p)\hat{L}'_{SR}G_{SR}(p)\hat{L}'_{SR} \, , \qquad (2.30)$$

.

.

.

Carrying out the inverse Laplace transformation of Eq. (2.27) yields

$$\frac{\partial}{\partial t}\sigma(t) = -i\hat{L}_S\sigma(t) \, - \, \int_0^t dy <\pi^{(0)}(y)>_R\sigma(t-y)$$

$$- \, \int_0^t dy[<\pi^{(1)}(y)>_R + <\pi^{(2)}(y)>_R + \cdots]\sigma(t-y) \, , \qquad (2.31)$$

where

$$\pi^{(n)}(y) = \frac{1}{2\pi i}\int_{-i\infty+c}^{i\infty+c} dp \, \pi^{(n)}(p)\exp(py) \, , \qquad (2.32)$$

and the convolution theorem has been used:

$$\frac{1}{2\pi i}\int_{-i\infty+c}^{i\infty+c} dp \, F(p)H(p)\exp(pt) = \int_0^t dy \, F(y)H(t-y) \, ,$$

$$\qquad (2.33)$$

where F(y) and F(p) satisfy Eq. (2.32). H(p) and H(t-y) are defined in an analogous way. By using the definition of displacement operator

$$\sigma(t-y) = \exp(-y\frac{\partial}{\partial t})\sigma(t) , \qquad (2.34)$$

we can express Eq. (2.31) as

$$\frac{\partial}{\partial t}\sigma(t) = -\hat{L}_S\sigma(t) - \int_0^t dy <\pi^{(0)}(y)>_R \exp(-y\frac{\partial}{\partial t})\sigma(t)$$

$$- \int_0^t dy [<\pi^{(1)}(y)>_R + <\pi^{(2)}(y)>_R + \cdots] \exp(-y\frac{\partial}{\partial t})\sigma(t) .$$

$$(2.35)$$

Equation (2.31) or (2.35) has a non-Markoffian structure. In other words, in order to determine the reduced density matrix (t) at $t = t_1$, we have to know time-dependence of the density matrix in the whole time range from the past (t = 0) to the present time ($t = t_1$).

In our treatment, we restrict ourselves to the Markoff processes. In this case, we obtain

$$\frac{\partial\sigma(t)}{\partial t} = -i\hat{L}_S\sigma(t) - \Gamma(t)\sigma(t) - W^{(1)}(t)\sigma(t) - W^{(2)}(t)\sigma(t) \cdots ,$$

$$(2.36)$$

where $W^{(n)}(t)$ represent the transition operators for the n-photon processes, $\Gamma(t) = W^{(n)}(t)$ with n = 0 and they are given by

$$W^{(n)}(t) = \int_0^t dy < \pi^{(n)}(y) >_R \exp[it\hat{L}_S] \ . \qquad (2.37)$$

We note that the population of the molecular eigenstate m at time t, $\sigma_{mm}(t)$ is defined in the Liouville space (see Eq. (2.3)) by $<<mm|\sigma(t)>>$. After projecting out the diagonal part of the both sides of Eq. (2.36) and taking time limit t → ∞ in the transition operators, we obtain the kinetic equation (called master equation) for the population change related to the resonant multiphoton transitions:

$$\frac{d\sigma_{mm}(t)}{dt} = \sum_{a \neq m} (k_{ma}^{(0)} + k_{ma}^{(1)} + \cdots)\sigma_{aa}(t) - (k_{mm}^{(0)} + k_{mm}^{(1)} + \cdots)\sigma_{mm}(t) \ ,$$
$$(2.38)$$

where $k_{ma}^{(n)}$, representing the n-photon transition rate constant from a to m molecular states, is given by

$$k_{ma}^{(n)} = \lim_{t \to \infty} W_{ma}^{(n)}(t) \ , \qquad (2.39a)$$

with

$$W_{ma}^{(n)}(t) = -<<mm|W^{(n)}(t)|aa>> \ , \qquad (2.39b)$$

and $k_{mm}^{(n)}$ represents the n-photon induced population decay constant of the state m, and is given in an analogous form as Eqs. (2.39). The rate equation approach based on Eq. (2.38) has been adopted in studying dynamics of molecules in the presence of the photon field such as UV and/or visible multiphoton ionization dissociation reactions, and vibrational relaxations in the resonant states.[9]

II-2.   The dephasing constant

In this section, an expression for the dephasing constant which makes an important contribution to the resonant multiphoton processes is presented, and the time evolution of the reduced density matrix after the coherence between the molecular eigenstates in the resonant state is created by irradiation of laser is considered within the isolated line approximation .

The equation of motion of the off-diagonal reduced density matrix element between m and n molecular eigenstates, $\sigma_{mn}(t) = <<mn|\sigma(t)>>$ in the absence of the photon field is given from Eq. (2.36) as

$$\frac{d}{dt}<<mn|\sigma(t)>> = \sum_{k} \sum_{\ell} <<mn|\{i\hat{L}_S-\Gamma(t)\}|k\ell>><<k\ell|\sigma(t)>> .$$

(2.40)

Real and imaginary parts of the diagonal matrix element of $\Gamma(t)$ in the Liouville space, for example, $<<mn|\Gamma(t)|mn>>$ ($\equiv \Gamma_{mn:mn}$) represent the time-dependent level shift and the dephasing rate between m and n states, respectively. The off-diagonal matrix element in the Liouville space, $\Gamma_{mn:k\ell}$ represents the contribution of overlap between two transitions,

$k \leftarrow \ell$ and $n \leftarrow m$.

The off-diagonal matrix element $\Gamma_{mn:k\ell}$ is given by

$$\Gamma_{mn:k\ell}(t) = \int_0^t dy <<mn| <<\pi^{(0)}(y)>_B>_R |k\ell>> , \qquad (2.41)$$

where

$$<<mn| <<\pi^{(0)}(y)>_B>_R |k\ell>> = Tr_B\{<<mn|\hat{L}'_{SB}(y)\exp[-iy\hat{L}_S]\hat{L}'_{SB}|k\ell>>$$

$$\times \ <<k\ell|\exp[iy\hat{L}_S]|k\ell>>\rho^{(B)}(0)\} , \qquad (2.42)$$

in which

$$\hat{L}'_{SB}(y) = \exp[iy\hat{L}_B]\hat{L}'_{SB} \exp[-iy\hat{L}_B] . \qquad (2.43)$$

In reducing Eq. (2.42), the following relation between the matrix elements of the Liouville operator, $\hat{L}$ and those of the Hamiltonian, $\hat{H}$, has been used:[17]

$$<<mn|\hat{L}|k\ell>>\hat{F} = \frac{1}{\hbar}(H_{mk}\delta_{n\ell}\hat{F}-\hat{F}H_{\ell n}\delta_{km})$$

$$= \frac{1}{\hbar}(H_{mk}\delta_{n\ell}-H^\times_{\ell n}\delta_{km})\hat{F} , \qquad (2.44)$$

in which $\hat{F}$ is an operator of a system coupled to the molecule, for example, the density matrix of the heat bath in the equilibrium for $\hat{H} = \hat{H}'_{SB}$. Subersript $\times$ of H indicates that $H^\times$ should be replaced with F. Equation (2.42) can be expressed as

$$<<mn|<<\pi^{(0)}(y)>_B>_R|k\ell>> = \frac{1}{\hbar^2}\sum_p \{\exp[-iy\omega_{pk}]<V_{mp}(y)V_{pk}>_B\delta_{n\ell}$$

$$+ \exp[-iy\omega_{\ell p}]<V_{\ell p}V_{pn}(y)>_B\delta_{mk}\}$$

$$- \frac{1}{\hbar^2}\{\exp[-iy\omega_{\ell n}]<V_{\ell n}V_{mk}(y)>_B + \exp[-iy\omega_{mk}]<V_{\ell n}(y)V_{mk}>_B\},$$

$$(2.45)$$

where ω_{mk} is the frequency difference between m and k levels, and

$$V_{mp}(y) = <m|\exp[iy\hat{H}_B/\hbar]\hat{H}'_{SB}\exp[-iy\hat{H}_B/\hbar]|p> . \qquad (2.46)$$

The diagonal matrix element of Eq. (2.42) can be written as

$$<<mn|<<\pi^{(0)}(y)>_B>_R|mn>> = \frac{1}{\hbar^2}\sum_{p\neq m}\{\exp[-iy\omega_{pm}]<V_{mp}(y)V_{pm}>_B$$

$$+ \exp[-iy\omega_{np}]<V_{np}V_{pn}(y)>_B\} + \frac{1}{\hbar^2}<V_{mm}(y)V_{mm}>_B$$

$$+ \frac{1}{\hbar^2}<V_{nn}V_{nn}(y)>_B - \frac{1}{\hbar^2}<V_{nn}V_{mm}(y)>_B - \frac{1}{\hbar^2}<V_{nn}(y)V_{mm}>_B .$$

$$(2.47)$$

The first two terms and the others in Eq. (2.47) originate from the dephasing effects due to inelastic and elastic molecule-heat bath interactions, respectively.

When the molecule is excited within its resonance width of the molecular eigenstate, we can take t as t → ∞ in calculating the matrix element of Γ(t). This is called the

Markoff approximation. In the Markoff approximation, the dephasing constant related to m and n levels, $\Gamma_{mn:mn}$, for example is given by[18]

$$\Gamma_{mn:mn} = \frac{1}{2}(\Gamma_{mm:mm} + \Gamma_{nn:nn}) + \Gamma^{(d)}_{mn:mn} \quad , \qquad (2.48)$$

where $\Gamma_{mm:mm}$ $(= k^{(0)}_{mm})$, and $\Gamma^{(d)}_{mn:mn}$ denote the population decay constant of the level m, and the pure dephasing constant resulting from the elastic interactions between the molecule and the heat bath, respectively, and they are expressed as

$$\Gamma_{mm:mm} = \frac{2\pi}{\hbar^2}\sum_{j}^{(j\neq m)}\sum_{\underline{j}}\sum_{\underline{m}} \rho^{(B)}_{\underline{mm}}(0)|(\hat{H}'_{SB})_{j\underline{j}, \ m\underline{m}}|^2 \delta(\omega_{mj} - \omega_{\underline{jm}}) \quad , \quad (2.49)$$

and

$$\Gamma^{(d)}_{mn:mn} = \frac{2\pi}{\hbar^2}\sum_{\underline{j}}\sum_{\underline{m}} \rho^{(B)}_{\underline{mm}}(0)|(\hat{H}'_{SB})_{m\underline{j}, m\underline{m}} - (\hat{H}'_{SB})_{n\underline{j}, n\underline{m}}|^2 \delta(\omega_{\underline{mj}}) \quad ,$$

$$(2.50)$$

where $\underline{m}$ and $\underline{j}$ indicate the initial and final levels of the perturber of the heat bath, respectively.

In the case in which two transitions, m ← n and k ← ℓ do not overlap each other, we can neglect <<mn|Γ(t)|kℓ>> for |mn>>≠|kℓ>> in the equation of motion of the reduced density matrix element, Eq. (2.40). This is called the isolated line approximation. In the isolated line approximation, the solution of Eq. (2.40) is given by

$$<<mn|\sigma(t)>> \; = \; <<mn|G_S(t)|mn>><<mn|\sigma(0)>> \; , \qquad (2.51)$$

or

$$<<mn|\sigma(t)>> \; = \; <<mn|G_S(t-t_1)mn>><<mn|\sigma(t_1)>> \; , \qquad (2.52)$$

where $G_S(t-t_1)$, the time evolution operator of the reduced density matrix, takes the form

$$G_s(t-t_1) \; = \; \exp[-\int_{t_1}^{t} dy\{i\hat{L}_S+\Gamma(y)\}] \; . \qquad (2.53)$$

The matrix element of Eq. (2.53) is expressed as

$$<<mn|G_S(t-t_1)|mn>> \; = \; \exp[-i(t-t_1)\omega_{mn}-\int_{t_1}^{t} dy\Gamma_{mn:mn}(y)] \; . \qquad (2.54)$$

In the Markoff approximation, Eq. (2.54) reduces to

$$<<mn|G_S(t-t_1)|mn>> \; = \; \exp[-(t-t_1)(i\omega_{mn}+\Gamma_{mn:mn})] \; . \qquad (2.55)$$

where the real part of the dephasing constant has been omitted. Equation (2.52) with Eq. (2.54) or (2.55) represents the time evolution of the off-diagonal density matrix of the m and n states, created by a coherent laser pulse at $t = t_1$, and that the damping is dominated by the dephasing rate (constant).

II-3. Coherence between molecular eigenstates

In this section, the time evolution of molecules produced by a coherent excitation will be treated in terms of the reduced density matrix formalism. Appearance of quantum beats in fluorescence of molecules are directly related to the time dependence of the off-diagonal density matrix elements.[17]

Let us first specify the molecule-photon interaction Hamiltonian, H'_{SR}. In the dipole and rotating wave approximations. the interaction Hamiltonian is given by

$$\hat{H}'_{SR} = -\sum_m \sum_n (M_{nm}(1)E^{(+)}|n\rangle\langle m|+h.c.) , \qquad (2.56)$$

where $M_{nm}(1)$ is defined in terms of the dipole matrix element between the molecular states, m and n, M_{nm} and the polarization of photon e_1 as $M_{nm}(1) = M_{nm} \cdot e_1$, and $E^{(+)}$ is the electric field operator of the photon field and is expressed as

$$E^{(+)} = -i(\frac{\hbar}{2\varepsilon_0 V})^{\frac{1}{2}}\sum_\lambda \omega_\lambda^{\frac{1}{2}}\hat{b}_\lambda^\dagger , \qquad (2.57)$$

in which V and $\hat{b}_\lambda^\dagger$ denote the volume of the photon field and the photon creation operator of mode λ with frequency ω_λ, respectively.

The equation of motion for the density matrix related to the one-photon absorption processes is given by Eq. (2.38) with n = 1:

$$\frac{\partial}{\partial t}\sigma(t) = -\{i\hat{L}_S+\Gamma(t)\}\sigma(t)-\int_0^t dt_1 <\pi^{(1)}(y)>_R \exp[it\hat{L}_S]\sigma(t) .$$

(2.58)

By using the perturbation method, we can obtain the solution as

$$\sigma(t) = \exp[-\int_0^t dt_1\{i\hat{L}_S+\Gamma(t)\}]\sigma(0)$$

$$- \int_0^t dt_1 \int_0^{t_1} dt_2\ G_S(t-t_1)<\hat{L}_{SR}(t_1-t_2)G_S(t-t_1)\hat{L}_{SR}>_R$$

$$\times \exp[i(t_1-t_2)\hat{L}_S]\sigma(0) .$$

(2.59)

We consider the time evolution of the off-diagonal reduced density matrix $\sigma_{mn}(t)$ in which molecular eigenstates (m and n) are coherently excited from the initial state a. The off-diagonal matrix element $\sigma_{mn}(t) = <<mn|\sigma(t)>>$ can be written as

$$<<mn|\sigma(t)>> = -\int_0^t dt_1 \int_0^{t_1} dt_2 \sum_k \sum_\ell <<mn|G_S(t-t_1)|mn>>$$

$$\times Tr_R\{<<mn|\hat{L}'_{SR}(t_1-t_2)|k\ell>><<k\ell|G_S(t-t_1)|k\ell>>$$

$$\times <<k\ell|\hat{L}'_{SR}|aa>>\rho^{(R)}(0)\} .$$

(2.60)

After the calculation of the matrix elements involving the Liouville operator $\hat{L}'_{SR}$ by using Eq. (2.44), we can express

Eq. (2.60) as

$$\sigma_{mn}(t) = \frac{M_{an}(1)M_{ma}(1)}{\hbar^2} \int_0^t dt_1 \int_0^{t_1} dt_2 [<E^{(+)}(t_1)E^{(-)}(t_2)>_R$$

$$\times <<mn|G_S(t-t_1)|mn>><<ma|G_S(t_1-t_2)|ma>>$$

$$+ <E^{(+)}(t_2)E^{(-)}(t_1)>_R<<mn|G_S(t-t_1)|mn>>$$

$$\times <<an|G(t_1-t_2)|an>>] , \qquad (2.61)$$

where $< \quad >_R$ represents the photon field correlation function.

The time dependence of the density matrix element related to the one-photon absorption processes can be described in terms of the diagramatic representation shown in Fig. 2.1. The diagram in the left-hand side and that in the right-hand side correspond to the first and second terms in Eq. (2.61), respectively. The lower and upper lines represent the time evolution of the ket $|>$ and bra $<|$ vectors of the molecule, respectively. Time develops from the left- to right-hand sides. The wavy line represents the molecule-photon interaction duration. The dotted points indicate the interaction points. The diagram in the left-hand side shows that the molecular coherence between the initial state a and the resonant state m is created and decays during the molecule-photon interaction time, t_1-t_2. Following the photon absorption by the molecules, the molecular coherence between two excited states, m and n begins to decrease because of the existence of dephasing processes. The time dependence is expressed by $<<mn|G_S(t-t_1)|mn>>$ which is given by Eq. (2.54) or Eq. (2.55) in the Markoff approximation.

So far we did not mention the definition of coherence. The degree of the coherence between the molecular eigenstates, m and n, C(m,n) is given in terms of the reduced density matrices as[17]

$$C(m,n) = [\frac{\sigma_{mn}(t)\sigma_{nm}(t)}{\sigma_{mm}(t)\sigma_{nn}(t)}]^{\frac{1}{2}} .$$

(2.62)

From the definition of density matrix, the degree of the coherence is smaller than or equal to one. C(m,n) nearly equal to one means that the molecular state which is represented in terms of a linear combination of the m and n states is formed, in other words, the coherence is created between m and n states.

An application of the theory developed in this section to the quantum beats in fluorescence is given in Sec. III-2.

II-4. Two-photon transitions via coherently excited
resonant states

Expressions for the two-photon transition rates are
derived by using the transition operator derived in Sec.
II-1. We focus our attention on the two-photon transition
via coherently excited resonant states. It is interesting
to investigate possibility of observing the interference
effects between molecular eigenstates of the resonant states
such as a non-Lorenzian band shape in the two-photon
ionization spectra or quantum beats in the time-resolved
spectra.

The two-photon transition rate from initial state a to
the final state f via resonant intermediate states, $W_{fa}^{(2)}(t)$
is given from Eq. (2.37) as

$$W_{fa}^{(2)}(t) = -<<ff|W^{(2)}(t)|aa>>$$

$$= \int_0^t dt_1 Tr_R\{<<ff|\pi^{(2)}(t-t_1)|aa>>\rho^{(R)}(0)\} , \qquad (2.63)$$

in which two-photon transition operator, $\pi^{(2)}$ can be
expressed as

$$\pi^{(2)}(t-t_1) = \frac{1}{2\pi i}\int_{-i\infty+c}^{i\infty+c} dp\{-\hat{L}'_{SR}G_{SR}(p)\hat{L}'_{SR}G_{SR}(p)\hat{L}'_{SR}G_{SR}(p)\hat{L}'_{SR}\}$$

$$\times \exp[p(t-t_1)] . \qquad (2.64)$$

By applying the convolution theorem to Eq. (2.64), substituting the resulting expression into Eq. (2.63) and changing the integration variables, we obtain

$$W_{fa}^{(2)}(t) = (\int_0^t dt_i) Tr_R \{ <<ff| \hat{L}'_{SR}(t_1) G_S(t_1-t_2) \hat{L}'_{SR}(t_2)$$

$$\times G_S(t_2-t_3) \hat{L}'_{SR}(t_3) G_S(t_3) \hat{L}'_{SR} \rho^{(R)}(0) |aa>> \} , \qquad (2.65)$$

where

$$(\int_0^t dt_i) = \int_0^t dt_1 \int_0^{t_1} dt_2 \int_0^{t_2} dt_3 .$$

$<<ff|\cdots|aa>>$ in Eq. (2.65) consists of three types of the sequence (product) of the matrix elements of the Liouville operator $\hat{L}'_{SR}$:

$$
\begin{array}{ccccc}
 & \hat{L}'_{SR} & \hat{L}'_{SR} & \hat{L}'_{SR} & \hat{L}'_{SR} \\
\end{array}
$$

A)   aa $\longrightarrow$ ma $\longrightarrow$ fa $\longrightarrow$ fm' $\longrightarrow$ ff ,

B)   aa $\longrightarrow$ ma $\longrightarrow$ mm' $\longrightarrow$ fm' $\longrightarrow$ ff ,

and   (2.66)

C)   aa $\longrightarrow$ ma $\longrightarrow$ mm' $\longrightarrow$ fm' $\longrightarrow$ ff .

Here three types of the sequence are denoted by A, B and C respectively. The sequences conjugate to Eq. (2.66) are denoted by $A^\dagger$, $B^\dagger$ and $C^\dagger$, respectively.

The two-photon transition rate consists of three types of the partial transition rate originating from $A(A^\dagger)$, $B(B^\dagger)$, and $C(C^\dagger)$ sequences, which are respectively indicated by $W^{(2A)}(t)$, $W^{(2B)}(t)$ and $W^{(2C)}(t)$:

$$W_{fa}^{(2)}(t) = 2\text{Re}\{W^{(2A)}(t)+W^{(2B)}(t)+W^{(2C)}(t)\} \ . \qquad (2.67)$$

Each partial transition rate is defined as

$$W^{(2A)}(t) = (\int_0^t dt_i)\sum_{m}\sum_{m'} \text{Tr}_R\{<<ff|\hat{L}'_{SR}(t_1)|fm'>>$$

$$\times \ <<fm'|G_S(t_1-t_2)|fm'>><<fm'|\hat{L}'_{SR}(t_2)|fa>>$$

$$\times \ <<fa|G_S(t_2-t_3)|fa>><<fa|\hat{L}'_{SR}(t_3)|ma>>$$

$$\times \ <<ma|G_S(t_3)|ma>><<ma|\hat{L}'_{SR}|aa>>\rho^{(R)}(0)\} \ , \qquad (2.68a)$$

$$W^{(2B)}(t) = (\int_0^t dt_i)\sum_{m}\sum_{m'} \text{Tr}_R\{<<ff|\hat{L}'_{SR}(t_1)|fm'>>$$

$$\times \ <<fm'|G_S(t_1-t_2)|fm'>><<fm'|\hat{L}'_{SR}(t_2)|mm'>>$$

$$\times \ <<mm'|G_S(t_2-t_3)|mm'>><<mm'|\hat{L}'_{SR}(t_3)|ma>>$$

$$\times \ <<ma|G_{SR}(t_3)|ma>><<ma|\hat{L}'_{SR}|aa>>\rho^{(R)}(0)\} \ , \qquad (2.68b)$$

and

$$W^{(2C)}(t) = (\int_0^t dt_i) \sum_{m} \sum_{m'} Tr_R \{ <<ff|\hat{L}'_{SR}(t_1)|mf>>$$

$$\times \ <<mf|G_S(t_1-t_2)|mf>><<mf|\hat{L}'_{SR}(t_2)|mm'>>$$

$$\times \ <<mm'|G_S(t_2-t_3)|mm'>><<mm'|\hat{L}'_{SR}(t_3)|ma>>$$

$$\times \ <<ma|G_S(t_3)|ma>><<ma|\hat{L}'_{SR}|aa>>\rho^{(R)}(0) \} \ . \qquad (2.68c)$$

We consider a two-photon transition induced by different two photon fields E_1 and E_2. In this case, after averaging over the photon field variables in Eqs. (2.68), we obtain

$$W^{(2A)}(t) = \sum_{m} \sum_{m'} (\int_0^t dt_i) <E_1^{(+)}(t_2)E_1^{(-)}(0)>_R <E_2^{(+)}(t_2)E_2^{(-)}(t_3)>_R$$

$$\times \ M_{am'}(1)M_{m'f}(2)M_{fm}(2)M_{ma}(1)<<fm'|G_S(t_1-t_2)|fm'>>$$

$$\times \ <<fa|G_S(t_2-t_3)|fa>><<ma|G_S(t_3)|ma>> \ , \qquad (2.69a)$$

$$W^{(2B)}(t) = \sum_{m} \sum_{m'} (\int_0^t dt_i) <E_1^{(+)}(t_3)E_1^{(-)}(0)>_R <E_2^{(+)}(t_1)E_2^{(-)}(t_2)>_R$$

$$\times \ M_{am'}(1)M_{m'f}(2)M_{fm}(2)M_{ma}(1)<<fm'|G_S(t_1-t_2)|fm'>>$$

$$\times \ <<mm'|G_S(t_2-t_3)|mm'>><<ma|G_S(t_3)|ma>> \ , \qquad (2.69b)$$

and

$$W^{(2C)}(t) = \sum_{m} \sum_{m'} (\int_0^t dt_i) <E_1^{(+)}(t_3) E_1^{(-)}(0)>_R <E_2^{(+)}(t_1) E_2^{(-)}(t_2)>_R$$

$$\times M_{am'}(1) M_{m'f}(2) M_{fm}(2) M_{ma}(1) <<mf|G_S(t_1-t_2)|mf>>$$

$$\times <<mm'|G_S(t_2-t_3)|mm'>> <<ma|G_S(t_3)|ma>> . \qquad (2.69c)$$

A diagramatic representation of the time evolution of the ket and bra vectors in two-photon transitions is shown in Figs. 2.2. Figures 2.2a, 2.2b and 2.2c correspond to the time evolution in Eqs. (2.69a), 2.69b and 2.69c, respectively. In the case of $m = m'$, that is, neglecting the coherent effects, these diagrams, for example, reduce to those for the resonant light scattering problems investigated by Takagahara et al,[19,20] and by Hanamura and Takagahara.[21]

So far, we have restricted ourselves to the two-photon transition. The rate of three-photon transitions can be taken into account in the similar way.

III. Application

III-1. Mechanisms of resonant multiphoton transitions

Terms "simultaneous process" and "sequential process" are frequently used in describing the multiphoton processes.[8,9] For the nonresonant case, the simultaneous process in general makes the dominant contribution. On the other hand, for the resonant case, the two mechanisms seem to operate on the multiphoton transition at the same time. The sequential transition is induced by two types of perturbations in the presence of laser light. One is inelastic perturbation such as intramolecular nonadiabatic interaction, or intermolecular interaction , which bring about electronic, vibrational and/or rotational relaxations. The other is an elastic interaction which causes so-called pure dephasing.

In this section, we first discuss the mechanism of a resonant two-photon transition by using the rate equation approach in which vibrational relaxations in the resonant states and the elastic interaction are taken into account. The structure of the two-photon transition probability $k_{fa}^{(2)}$ is analyzed on the basis of the formalism developed in Sec. II. The sequential transition induced by the pure dephasing process is derived. A difference in the vibronic intensity distribution between sequential and simultaneous processes is presented. Dephasing effects in three-photon absorptions are finally considered.

III-1A. Effects of vibrational relaxations for a resonant
         two-photon transition

We consider a model for a resonant two-photon transition shown in Fig. 3.1. Saturation effects are neglected for simplicity. By using Eq. (2.38), we can express the kinetic equations as[22]

$$\frac{d\sigma_{aa}(t)}{dt} = -k_{aa}\sigma_{aa}(t) \; , \qquad\qquad\qquad (3.1)$$

$$\frac{d\sigma_{mm}(t)}{dt} = \sum_a k_{ma}^{(1)}\sigma_{aa}(t) - k_{mm}\sigma_{mm}(t) - \sum_{m'}' k_{m'm}^{(0)}\sigma_{mm}(t)$$

$$+ \sum_{m'}' k_{mm'}^{(0)}\sigma_{m'm'}(t) \; , \qquad\qquad (3.2)$$

and

$$\frac{d\sigma_{ff}(t)}{dt} = \sum_a k_{fa}^{(2)}\sigma_{aa}(t) + \sum_m k_{fm}^{(1)}\sigma_{mm}(t) \; , \qquad (3.3)$$

where $k_{aa} = k_{aa}^{(1)}+k_{aa}^{(2)}$, and $k_{mm} = \sum_n k_{nm}^{(1)}$. Defining the matrix elements of R as

$$R_{mm'} = \delta_{mm'}\sum_\ell k_{\ell m}^{(0)}(1-\delta_{\ell m'}) - k_{mm'}^{(0)}(1-\delta_{mm'}) \; , \qquad (3,4)$$

we can rewrite Eq. (3.2) as

$$\frac{d\sigma_{mm}(t)}{dt} = \sum_a k_{ma}^{(1)}\sigma_{aa}(t) - k_{mm}\sigma_{mm}(t) - \sum_{m'} R_{mm'}\sigma_{m'm'}(t) \; . \; (3.5)$$

Symmetrizing the relaxation matrix R by setting $\sigma_{mm}(t) = \tilde{\sigma}_{mm}(t)f_m^{1/2}$, in which f_m denotes the equilibrium vibrational distribution in the resonant state, we obtain the kinetic equation for $\tilde{\sigma}_{mm}(t)$ in the matrix formalism as

$$\frac{d\tilde{\sigma}(t)}{dt} = \tilde{C}(t) - (K+\tilde{R})\tilde{\sigma}(t) , \qquad (3.5)$$

where $\tilde{C}_m(t) = \sum_a \tilde{k}_{ma}^{(1)} \sigma_{aa}(t)$ with $\tilde{k}_{ma}^{(1)} = k_{ma}^{(1)} f_m^{-1/2}$ and $\tilde{R}_{mm'} = R_{mm'}(f_{m'}/f_m)^{1/2}$. The solution of Eq. (3.5) is expressed as

$$\sigma_{mm}(t) = \sum_a \sum_j \sum_\ell (f_m/f_\ell)^{\frac{1}{2}} N_{mj} \frac{1-\exp[-(\lambda_{jj}-k_{aa})t]}{\lambda_{jj}-k_{aa}}$$

$$\times (N^{-1})_{j\ell} k_{\ell a}^{(1)} \exp(-k_{aa}t)\sigma_{aa}(0) , \qquad (3.6)$$

in which N_{mj} and λ_{jj} satisfy the equation

$$(K + \tilde{R}) N = N\lambda . \qquad (3.7)$$

Substituting Eq. (3.6) into Eq. (3.3) yields the expression for the two-photon transition rate, $J^{(2)}(t)$ after taking summation over the final vibronic levels:

$$J^{(2)}(t) = \sum_f \frac{d\sigma_{ff}(t)}{dt}$$

$$= \sum_f \sum_a [k_{fa}^{(2)} + \sum_m \sum_\ell k_{fm}^{(1)} \lambda_{m\ell}(t)k_{\ell a}^{(1)}]\exp(-k_{aa}t)\sigma_{aa}(0) , \quad (3.8)$$

where the first term involving the simultaneous process, sequential one induced by elastic molecule-bath interactions and their mixing term will be discussed in a subsequent section, and the second term expresses the sequential process induced by vibrational relaxations:

$$
\begin{array}{ccc}
\text{photon} & \text{vibrational} & \text{photon} \\
\text{absorption} & \text{relaxation} & \text{absorption}
\end{array}
$$

$$
a \xrightarrow{\hspace{2cm}} \ell \;\rightsquigarrow\; m \xrightarrow{\hspace{2cm}} f \,. \tag{3.9}
$$

The matrix element, $\lambda_{m\ell}(t)$ is given by

$$
\lambda_{m\ell}(t) = (f_m/f_\ell)^{\frac{1}{2}} \sum_j N_{mj} \frac{1-\exp[-(\lambda_{jj}-k_{aa})t]}{\lambda_{jj}} (N^{-1})_{j\ell} \,. \tag{3.10}
$$

For the steady state in the resonant state, Eq. (3.10) takes the form

$$
\lambda_{m\ell}(t \to \infty) = (f_m/f_\ell)^{\frac{1}{2}} \sum_j N_{mj} \lambda_{jj}^{-1} (N^{-1})_{j\ell} \,. \tag{3.11}
$$

For the time scale of the pulse in which the depopulation of the initial state can be neglected, the transition rate constant in the steady state condition can be written as

$$
J^{(2)} = \sum_f \sum_a [k_{fa}^{(2)} + \sum_m \sum_\ell k_{fm}^{(1)} \lambda_m (t \to \infty) k_{\ell a}^{(1)}] \sigma_{aa}(0) \,. \tag{3.12}
$$

Effects of the vibrational relaxations on a resonant two-photon ionization of molecules have been discussed elsewhere.[22] Slow and fast vibrational relaxation cases, compared with the transition rate constant from the resonant to the ionization states, are taken into account.

For the extreme situation in which no vibrational relaxation exists, noting that K is a diagonal matrix and $N_{mj} = \delta_{mj}$, and further for the case in which $\lambda_{jj} > k_{aa}$, we obtain

$$J^{(2)}(t) = \sum_f \sum_a [k_{fa}^{(2)} + \sum_m k_{ma}^{(1)}] \exp(-k_{aa}t)\sigma_{aa}(0) . \quad (3.11)$$

The apparent transition rate constant $k_{ma}^{(1)}$ for the resonant two-photon transition originates from the fact that the transition rate from the resonant to the final state is faster than the excitation rate to the resonant state, and is the rate constant of the rate-determing step if the magnitude of $k_{fa}^{(2)}$ is negligibly small.

Murakami et al.[23] have observed the two-photon resonant four-photon ionization of benzene and the halogen-substituted benzenes. They have concluded that the resonant multiphoton ionization spectra reflect the two-photon absorption spectra of $S_1 \leftarrow S_0$ transition. This suggests that the rate from the ground to the resonant states mainly contributes to the multiphoton process. This can be interpreted in terms of the expression simillar to Eq. (3.11).

III-1B. Structure of two-photon absorption rate constants, $k_{fa}^{(2)}$

In this section, the structure of the two-photon absorption rate constant $k_{fa}^{(2)}$ is discussed. This has the structure similar to that for resonant light scattering in which both simultaneous (coherent) and sequential (incoherent) processes participate.[24,25] It is well known that the latter process in the resonant vibrational light scattering is induced by pure dephasing processes related to the initial and resonant states,[26-30] and the simultaneous and sequential processes can be separated out in the case of the negligibly small dephasing constant between the initial and final states.[31]

It should be noted that there exists a difference in the contribution of the dephasings between the resonant vibrational light scattering and the resonant two-photon transitions (absorption, ionization, etc.). For the former, the dephasing constant between the initial (final) and resonant states, which may be called electronic dephasing constant, and that between the initial and final states, called vibrational dephasing constant, make contribution. For the resonant two-photon transition, on the other hand, both dephasing constant between the initial (final) and resonant states and that between the initial and final states belong to the electronic dephasing constant. When the magnitutes of these dephasing constants are of the same order, the coherent and incoherent processes can not be separated out; a mixing process between two processes participates in the resonant two-photon processes.[24,31] It is commonly accepted that the magnitude of the vibrational dephasing constant is negligibly small compared with that of the electronic dephasing constant.

From Eqs. (2.39) and (2.67), the two-photon transition
rate constant is given by

$$k_{fa}^{(2)} = 2 \lim_{t\to\infty} \text{Re}\{W^{(2A)}(t)+W^{(2B)}(t)+W^{(2C)}(t)\} \quad , \quad (3.12)$$

where $W^{(2A)}(t)$, $W^{(2B)}(t)$, and $W^{(2C)}(t)$ are defined by
Eqs. (2.69). In calculating Eq. (3.12), photon-field
correlation functions with constant pulse amplitudes with
E_1 and E_2 for the first and second lasers neglecting the
band widths are considered:

$$<E_1^{(+)}(t_i)E_1^{(-)}(t_j)>_R = E_1^2\exp[i(t_i-t_j)\omega_1] \quad , \quad (3.13a)$$

and

$$<E_2^{(+)}(t_i)E_2^{(-)}(t_j)>_R = E_2^2\exp[i(t_i-t_j)\omega_2] \quad . \quad (3.13b)$$

After substituting Eqs. (3.13) into Eqs. (2.69),
carrying out the integration of the resulting expression
over the time variables, and taking the limit $t \to \infty$, the
two-photon transition rate constant averaged over the
initial vibrational distribution with σ_{aa}, $k^{(2)}$ is expressed
as

$$k^{(2)} = \sum_a \sum_f \sigma_{aa} k_{fa}^{(2)}$$

$$= \frac{2}{\hbar^4} E_1^2 E_2^2 \, \text{Re} \sum_a \sum_f \sigma_{aa} \sum_m \sum_{m'} M_{am'}(1) M_{m'f}(2) M_{fm}(2) M_{ma}(1)$$

$$\times \, [\frac{1}{(i\Delta\omega_{fa} + \Gamma_{fa})(i\Delta\omega_{fm'} + \Gamma_{fm'})(i\Delta\omega_{ma} + \Gamma_{ma})}$$

$$+ \frac{1}{(i\Delta\omega_{ma} + \Gamma_{ma})(i\Delta\omega_{mf} + \Gamma_{mf})(i\Delta\omega_{fm'} + \Gamma_{fm'})}$$

$$\times \, (1 + \frac{\Gamma_{fm'} + \Gamma_{mf} - \Gamma_{mm'}}{i\Delta\omega_{mm'} + \Gamma_{mm'}})] \, , \qquad (3.14)$$

where $\Delta\omega_{fa} = \omega_{fa} - \omega_1 - \omega_2$, $\Delta\omega_{ma} = \omega_{ma} - \omega_1$, and $\Delta\omega_{fm'} = \omega_{fm'} - \omega_2$. Equation (3.14) can be classified as

$$k^{(2)} = k^{(2-sim)} + k^{(2-seq)} + k^{(2-mix)} \, , \qquad (3.15)$$

where $k^{(2-sim)}$ represents the transition probability of the simultaneous process, and is given by

$$k^{(2-sim)} = \frac{2}{\hbar^4} E_1^2 E_2^2 \, \text{Re} \sum_a \sum_f \frac{\sigma_{aa}}{i\Delta\omega_{fa} + \Gamma_{fa}} \left| \sum_m \frac{M_{fm}(2) M_{ma}(1)}{i\Delta\omega_{ma} + \Gamma_{ma}} \right|^2 \, . \qquad (3.16)$$

$k^{(2-seq)}$, representing the transition probability of the sequential process, is expressed as

$$k^{(2-seq)} = \frac{2}{\hbar^4} E_1^2 E_2^2 \, \text{Re} \sum_a \sum_f \sum_m \sum_{m'} (\Gamma_{fm'} + \Gamma_{mf} - \Gamma_{mm'})$$

$$\times \, \frac{M_{am'}(1) M_{m'f}(2) M_{fm}(2) M_{ma}(1)}{(i\Delta\omega_{ma} + \Gamma_{ma})(i\Delta\omega_{mf} + \Gamma_{mf})(i\Delta\omega_{fm'} + \Gamma_{fm'})(i\Delta\omega_{mm'} + \Gamma_{mm'})}$$

$$(3.17)$$

and $k^{(2\text{-mix})}$ representing the transition probability of their mixing process is given by

$$k^{(2\text{-mix})} = \frac{2}{\hbar^4}E_1^2 E_2^2 \, \text{Re} \sum_a \sum_f \frac{\sigma_{aa}}{i\Delta\omega_{fa}+\Gamma_{fa}} \sum_m \sum_{m'}$$

$$\times \{(\Gamma_{fa}+\Gamma_{m'a}-\Gamma_{fm'})(i\Delta\omega_{mf}+\Gamma_{mf})+(i\Delta\omega_{fa}+\Gamma_{fa})$$

$$\times [-i(\Delta\omega_{mf}+\Delta\omega_{m'a})-\Gamma_{mf}+\Gamma_{m'a}]\}$$

$$\times \frac{M_{am'}(1)M_{m'f}(2)M_{fm}(2)M_{ma}(1)}{(i\Delta\omega_{ma}+\Gamma_{ma})(i\Delta\omega_{mf}+\Gamma_{mf})(i\Delta\omega_{fm'}+\Gamma_{fm'})(-i\Delta\omega_{m'a}+\Gamma_{m'a})} \; .$$

$$(3.18)$$

Effects of laser bandwidths were neglected in the treatment described above. Recently a finite laser bandwidth has been taken into account to discuss the mechanism of the resonant two-photon transitions. If the photon field correlation function $< \quad >_R$ is expressed as[32]

$$\langle E^{(+)}(t_i)E^{(-)}(t_j)\rangle_R = E^2\exp[i(t_i-t_j)\omega-|t_i-t_j|\delta\omega] \; ,$$

$$(3.19)$$

where $\delta\omega$ denotes the laser bandwidth, it can easily been shown that the expression for the rate constant is given from Eq. (3.14) by replacing $\Gamma_{fa} \to \Gamma_{fa}+\delta\omega_1+\delta\omega_2$, $\Gamma_{ma} \to \Gamma_{ma}+\delta\omega_1$, $\Gamma_{fm'} \to \Gamma_{fm'}+\delta\omega_2$, and so on.

III-1C. Vibronic intensity distributions in resonant
         two-photon absorptions

One of the resonance effects is revealed in the
appearance of vibrational structures in the multiphoton
spectra.[9] This is related to the difference among the
potential energy surfaces involved in the multiphoton
transition, and can be explained by using the Franck-Condon
principle. For example, potential displacements between
the initial and resonant states and/or between the resonance
and final states as well as those between the initial and
final electronic states reflect on the two-photon absorption
spectra. This indicates that there exists a difference in
the vibronic intensity distribution between the simultaneous
and sequential transitions.

In this subsection, analytical expressions for the
transition probability of the simultaneous and sequential
two-photon absorption processes for a multilevel molecular
system are presented within the displaced harmonic oscillator
model.[24]

Let us first consider the simultaneous process shown
in Fig. 3.2. The simultaneous two-photon transition
probability, $k^{(2-sim)}$ is given by Eq. (3.16). In the Born-
Oppenheimer approximation, the molecular state a is
expressed as $|a> = |\Phi_a>|\Theta_a>$, where $|\Phi_a>$ and $|\Theta_a>$ represent
electronic and vibrational eigenvectors of the ath electronic
state, respectively. Equation (3.16) can be expressed in

the generating function form as

$$k^{(2-sim)} = \frac{E_1^2 E_2^2}{\hbar^4} |M_{fm}^0 M_{ma}^0|^2 \int_{-\infty}^{\infty} dt \int_0^{\infty} d\tau \int_0^{\infty} d\tau' \exp[-\Gamma_{fa}|t|$$

$$- \Gamma_{ma}(\tau+\tau')-it(\omega_1+\omega_2)-i(\tau'-\tau)\omega_1]G(\tau,\tau',t) . \quad (3.20)$$

$G(\tau,\tau',t)$, the generating function for the simultaneous two-photon transition, is given by

$$G(\tau,\tau',t) = \sum_a <\Theta_a|\sigma_{aa}\exp[-i(t+\tau'-\tau)\hat{H}_a/\hbar]$$

$$\times \exp[i\hat{H}_m\tau'/\hbar]\exp[i\hat{H}_f t/\hbar]\exp[-i\hat{H}_m\tau/\hbar]|\Theta_a> , \quad (3.21)$$

where $\hat{H}$ represent the vibronic Hamiltonians of the relevant states. In deriving Eq. (3.20), polarization effects were neglected, and the Condon approximation was used: $M_{fm}^0 = <\Phi_f^0|e\mathbf{r}|\Phi_m^0>$ where $\Phi_m^0(\Phi_f^0)$ is the electronic wave function for the resonant (final) state, evaluated at the equilibrium position in the electronic ground state. For simplicity, let us consider a displaced harmonic oscillator model in which the vibronic Hamiltonians of the initial, resonant, and final electronic states are, respectively, written as

$$\hat{H}_a = \frac{\hbar\omega}{2}(\hat{p}^2+q^2) , \quad (3.22a)$$

$$\hat{H}_m = \frac{\hbar\omega}{2}[\hat{p}^2+(q-\Delta_{ma})^2] + \varepsilon_m^0 , \quad (3.22b)$$

and

$$\hat{H}_f = \frac{\hbar\omega}{2}[\hat{p}^2 + (q - \Delta_{fa})^2] + \varepsilon_f^0 , \qquad (3.22c)$$

where ω is the harmonic frequency, q and $\hat{p}$ are the dimensionless nuclear coordinate and the conjugated momentum. Δ_{ma} represents the dimensionless displacement between the equilibrium points in the resonant and ground electronic states, and ε_m^0 is the electronic energy gap between the bottoms in the two electronic states. In this model, Eq. (3.20) can be expressed in the analytical form as

$$k^{(2-sim)} = \frac{2E_1^2 E_2^2}{\hbar^4} |M_{fm}^0 M_{ma}^0|^2 \exp[-(\bar{n} + \tfrac{1}{2})(\Delta_{ma}^2 + \Delta_{fm}^2)]$$

$$\times \sum_{k=0}^{\infty} \sum_{\ell=0}^{\infty} \frac{(\bar{n}+1)^k \bar{n}^\ell}{k!\,\ell!} \frac{\Gamma_{fa}}{\{\omega_1 + \omega_2 - \frac{\varepsilon_f^0}{\hbar} - (k-\ell)\omega\}^2 + \Gamma_{fa}^2}$$

$$\times |\sum_{p=0}^{k} \sum_{q=0}^{\ell} \sum_{r=0}^{\infty} \sum_{s=0}^{\infty} \binom{k}{p}\binom{\ell}{q} \{-(\bar{n}+1)\frac{\Delta_{fm}\Delta_{ma}}{2}\}^r$$

$$\times (-\bar{n}\frac{\Delta_{fm}\Delta_{ma}}{2})^s (\frac{\Delta_{fm}}{\sqrt{2}})^{k-p+\ell-q} (\frac{\Delta_{ma}}{\sqrt{2}})^{p+q} \frac{1}{r!s!}$$

$$\times \frac{1}{i\{\frac{\varepsilon_m^0}{\hbar} - \omega_1 + (p-q+r-s)\omega\} + \Gamma_{ma}} |^2 . \qquad (3.23)$$

In the low temperature limit for the molecules, Eq. (3.23) reduces to

$$k^{(2-sim)} = \frac{2E_1^2 E_2^2}{\hbar^4} |M_{fm}^0 M_{ma}^0|^2 \exp[-\frac{1}{2}(\Delta_{ma}^2 + \Delta_{fm}^2)]$$

$$\times \sum_{\ell=0}^{\infty} \frac{\Gamma_{fa}}{\ell! \{(\epsilon_f^0/\hbar + \ell\omega - \omega_1 - \omega_2)^2 + \Gamma_{fa}^2\}}$$

$$\times |\sum_{j=0}^{\ell} \binom{\ell}{j} (\frac{\Delta_{fm}}{\sqrt{2}})^j (\frac{\Delta_{ma}}{\sqrt{2}})^{\ell-j}$$

$$\times \sum_{k=0}^{\infty} \frac{(-1)^k (\Delta_{fm}\Delta_{ma}/2)^k}{k! \{i(\epsilon_m^0/\hbar - \omega_1 - j\omega + \ell\omega + k\omega) + \Gamma_{ma}\}}|^2 . \qquad (3.24)$$

An expression for the vibronic intensity distribution of the sequential two-photon absorption in the Born-Oppenheimer approximation can be derived from Eq. (3.17). For simplicity, the following approximation is adopted, $(i\omega_{mm'} + \Gamma_{mm'})^{-1} \simeq \delta_{mm'}/\Gamma_{mm'}$, which means neglect of interference effects between the vibronic levels in the resonant state. In this approximation, Eq. (3.17) can be written as

$$k^{(2-seq)} \propto \frac{2\Gamma_{mf} - \Gamma_{mm}}{\Gamma_{mm}\Gamma_{mf}} \sum_m I_{a\rightarrow m}(\omega_1) I_{m\rightarrow f}(\omega_2) , \qquad (3.25)$$

where $I_{a\rightarrow m}(\omega_1)$ and $I_{m\rightarrow f}(\omega_2)$ represent the line shape function of the absorption from the electronic ground state to a single vibronic level m in the resonant state and that from this level to the final electronic state, and they are given by

$$I_{a\rightarrow m}(\omega_1) = \sum_a \sigma_{aa} \frac{\Gamma_{ma} |<\Theta_a | \Theta_m>|^2}{\Delta\omega_{ma}^2 + \Gamma_{ma}^2} , \qquad (3.26a)$$

and

$$I_{m \to f}(\omega_2) = \sum_f \frac{\Gamma_{mf} |<\Theta_m|\Theta_f>|^2}{\Delta\omega_{mf}^2 + \Gamma_{mf}^2} \quad , \qquad (3.26b)$$

respectively.

In a displaced harmonic oscillator model, an analytical expression in the low temperature limit can be obtained as

$$k^{(2-seq)} \propto \frac{(2\Gamma_{mf} - \Gamma_{mm})\Gamma_{ma}}{\Gamma_{mm}} \sum_{k=0}^{\infty} \frac{(\frac{\Delta_{ma}^2}{2})^k \exp[-\frac{\Delta_{ma}^2}{2}]}{k! \{(\varepsilon_m^0/\hbar + k\omega - \omega_1)^2 + \Gamma_{ma}^2\}}$$

$$\times \sum_{j=0}^{k} \frac{k!}{(k-j)!(j!)^2} (\frac{\Delta_{ma}^2}{2})^j \exp[-\frac{\Delta_{fm}^2}{2}]$$

$$\times \sum_{p=0}^{2j} \binom{2j}{p}(-1)^p \sum_{\ell=0}^{\infty} \frac{(\frac{\Delta_{fm}^2}{2})^\ell}{\ell! [\{(\varepsilon_f^0 - \varepsilon_m^0)/\hbar - \omega_2 + j\omega - p\omega + \ell\omega\}^2 + \Gamma_{mf}^2]} \quad .$$

$$(3.26c)$$

To illustrate the difference in the vibronic intensity distribution between the simultaneous and sequential two-photon transition, model calculations of the vibronic structures are performed by using the analytical expressions derived above. In Figs. 3. 3 and 3.4 , the vibronic intensity distributions in the case of excitation of the detuning frequency $\omega_1 - \varepsilon_m^0/\hbar = -200$ cm^{-1} are shown as a function of $\omega_2 - (\varepsilon_f^0 - \varepsilon_m^0)/\hbar$. The dimensionless displacement between the resonant and final state, Δ_{fm} was taken to be 2.0 and 0 for Figs. 3.3 and 3.4 , respectively, and $\Delta_{ma} = 1.6$. The dephasing constant Γ_{ma} = 5.0, Γ_{fa} = 5.0, and Γ_{fm} = 10.0, and the population decay constant Γ_{mm} = 10.0 were used. The dual structure in these figures reflects the simultaneous and sequential mechanisms: the bands whose peak locates at the integral multiple on the abscissa correspond to those originated from the sequential two-photon transition, and the other bands correspond to those originated from the simultaneous transition. The vibronic intensity distribution of the sequential transition depends on the sum of the transition probabilities from the vibronic levels in the resonant state to the final vibronic levels. On the other hand, the vibronic intensity distribution of the simultaneous transition is not expressed by a simple combination of the optical transitions such as that of the sequential one.

III-1D.   Effects of the dephasings in three-photon absorptions

In this subsection, effects of the dephasings, especially pure electronic dephasing on the resonant three-photon absorption are described, and it is shown that sequential three-photon transitions can be induced by the pure electronic dephasing.

The three-photon transition probability can be expressed in terms of ten different time-evolution diagrams. The transition probability averaged over the initial vibrational distribution $k^{(3)}$ can be written as

$$k^{(3)} \propto \mathrm{Re} \sum_a \sum_f \sigma_{aa} \sum_k \sum_{k'} \sum_m \sum_{m'} M_{am'}(1) M_{k'm'}(2) M_{m'f}(3) M_{fm}(3) M_{mk}(2)$$

$$\times M_{ka}(1) [G_{af} G_{ak'} G_{am'} G_{mf} G_{kf}$$

$$+ G_{mf}(G_{ka} + G_{ak'})(G_{km'} + G_{mk'}) G_{mm'} G_{kk'}$$

$$+ G_{mf}(G_{ka} G_{mk'} G_{ma} + G_{ak'} G_{am'} G_{km'}) G_{mm'}$$

$$+ G_{mf} G_{kf} G_{km'}(G_{ka} + G_{ak'}) G_{kk'}$$

$$+ G_{mf} G_{ak'} G_{kf} G_{am'} G_{km'}] , \qquad (3.27)$$

where k and k', and m and m' specify the vibronic levels
in the same electronic states, respectively.

$$G_{fa} = \{ i(\omega_{fa} - 3\omega_R) + \Gamma_{fa} \}^{-1} , \qquad (3.28)$$

and the other matrix elements of G can be expressed in the
similar way. Here it has been assumed that the molecules
have been excited by three photons with an identical
frequency, ω_R.

For a nonresonant three-photon absorption, Eq. (3.27)
can approximately be expressed as

$$k^{(3-sim)} \propto \mathrm{Re} \sum_a \sum_f \sigma_{aa} G_{af} | \sum_k \sum_m M_{ak}(1) G_{ak} M_{km}(2) G_{am} M_{mf}(3) |^2 ,$$
$$\qquad (3.29)$$

which can be assigned to the transition probability for the simultaneous three-photon transition.

For a detailed discussion of the dephasing effects a three-photon absorption in which resonance occurs on the absorption of the second photon, so-called (2+1) process shown in Fig. 3.5 will be considered. This process can be seen in multiphoton transitions of molecules such as iodine,[34-36] nitric oxide,[37] trans 1-3-butadiene,[38] benzene,[39] van der Waals complexes,[40] etc. Setting m = m' in Eq. (3.27) yields an approximate expression for the transition probability appropriate for the (2+1) process:

$$k^{(2+1)} \simeq k^{(2+1,sim)} + k^{(2+1,seq)} + k^{(2+1,mix)} \quad , \quad (3.30)$$

where

$$k^{(2+1,sim)} \propto \text{Re} \sum_a \sum_f \sigma_{aa} G_{af} \sum_m |M_{fm}(3)G_{am}|^2 \left| \sum_k M_{ak}(1)G_{ak}M_{km}(2) \right|^2 \quad , \tag{3.31}$$

$$k^{(2+1,seq)} \propto \sum_a \sum_f \sigma_{aa} \sum_m \frac{2\Gamma_{ma}^{(d)}+\Gamma_{aa}}{\Gamma_{mm}\Gamma_{ma}} |M_{fm}(3)|^2 \text{Re } G_{mf} \text{Re } G_{am}$$

$$\times \left| \sum_k M_{ak}(1)G_{ak}M_{km}(2) \right|^2 \quad , \tag{3.32}$$

$$k^{(2+1,mix)} \propto \sum_a \sum_f \sigma_{aa} \sum_m |M_{fm}(3)|^2 (\Gamma_{fa}+\Gamma_{ma}-\Gamma_{fm})$$

$$\times \text{Re } G_{af}G_{mf}|G_{am}|^2 \left| \sum_k M_{ak}(1)G_{ak}M_{km}(2) \right|^2 \quad . \tag{3.33}$$

Equation (3.31) represents the simultaneous (2+1) process.
Equation (3.32) represents the sequential (2+1) process
which mainly originates from an electronic pure dephasing
between the initial and resonant states because the third
photon is absorbed from the resonant state disturbed by
the elastic interaction between the molecule and heat bath
during the two-photon absorption. Γ_{aa} may make no any
significant contribution. Equation (3.33) describes mixing
effects of the processes expressed in terms of Eqs. (3.31)
and (3.32). In the case in which the condition among the
dephasing constants, $\Gamma_{fm} \simeq \Gamma_{ma} > \Gamma_{af}$ is satisfied, the
mixing term can be omitted compared with $k^{(2+1,sim)}$ and
$k^{(2+1, seq)}$.

In a previous paper,[33] we have discussed about the
possibility of identifying the vibronic bands characterized
by the simultaneous or sequential processes in the resonant
(2+1) multiphoton absorption spectrum using a single laser
beam of its frequency ω_R. The band peaks appearing in the
three-photon absorption spectrum as a function of ω_R (or
$2\omega_R$, $3\omega_R$) are mainly determined by the following resonance
conditions: $\Delta\omega_{af} = 3\omega_R - \omega_{fa} = 0$ and $\Delta\omega_{mf} = \omega_R - \omega_{mf} = 0$ for
the simultaneous and sequential processes, respectively,
in addition to the resonance condition of the two-photon
absorption, $\Delta\omega_{am} = 2\omega_R - \omega_{am} = 0$ for both processes.

To demonstrate the spectral features reflected in the
(2+1) absorption spectrum, a model calculation of the
vibronic intensity distributions has been carried out within

a displaced harmonic oscillator approximation. Analytical expressions for the transition probabilities of the (2+1) processes have been derived and used.[33] A result of the model calculation is shown in Fig. 3.6 in which vibronic bands due to the simultaneous and sequential (2+1) processes are denoted by a and b , respectively. The abscissa is expressed in terms of $2\omega_R$ and $3\omega_R$. The molecular parameters used are $\Delta_{ka} = 0$, $\Delta_{fm} = 0$, and $\Delta_{mk} = \Delta_{ma} = 0.1$ for the dimensionless displacements, $\omega = 500$ cm^{-1} for the molecular vibrational frequency, and $\omega_{fa} = 72180$, $\omega_{ma}^0 = 72180$, $\omega_{ma}^0 = 48000$, and $\omega_{ka}^0 = 32000$ cm^{-1} for the electronic energy gaps. The dephasing constants are $\Gamma_{aa} = 0$, $\Gamma_{kk} = 0$, $\Gamma_{ff} = 0.2$ and $\Gamma_{mm} = 2.0$ cm^{-1}, $\Gamma_{ma}^{(d)} = 10.0$, $\Gamma_{fm}^{(d)} = 10.0$, $\Gamma_{fm}^{(d)} = 10.0$, $\Gamma_{ka}^{(d)} = 0$, and $\Gamma_{fa}^{(d)} = 2.0$ cm^{-1}. One of the prominent features in the spectrum is the existence of the resonance band located at $2\omega_R = 48360$ cm^{-1}. This band can be assigned to that originating from the sequential process because the third photon whose frequency of $\omega_R = \omega_{fm} = 24180$ cm^{-1} is absorbed from the relaxed vibronic level in the resonant state m to the final state f as shown in Fig. 3.7.

III-2. Dephasing effects in quantum beats in molecular
       fluorescence

Quantum beats in the fluorescence of molecules excited by a coherent pulse laser have recently been reported by many authors.[41-48] By analyzing the quantum beats in the fluorescence one can investigate the magnitude of the interaction between the molecule and a perturber of the heat bath as well as the mechanism of the inter/intra electronic state couplings in electronically excited states.[49] For molecules belonging to the intermediate case, effects of collisional dephasing make an important contribution to the time-dependent behavior of the excited states. In this section, the collision effects resulting from both elastic and inelastic scattering processes between the molecule and perturber are theoretically studied based on the expressions of the reduced matrix elements derived in Sec. II. The dephasing effects are taken into account in the Markoff approximation. A non-Markoff theory of quantum beats will be presented elsewhere.[50]

The fluorescence intensity at time t, I(t) from m and n levels after a coherent excitation is given by

$$I(t) = K \sum_f \sum_m |M_{mf}|^2 \sigma_{mm}(t) + K \sum_f \sum_m \sum_n M_{nf} M_{fm} \sigma_{mn}(t) , \qquad (3.34)$$
$$(m \neq n)$$

where K is a constant involving the frequency of the emitted photon, $\sigma(t)$ whose matrix elements are given by Eq. (2.61) is the molecular density matrix representing the time evolution of the excited states, and f specifies the final molecular states, respectively.

In order to investigate the dephasing effects, a model calculation of the quantum beats is carried out in the two-level model for the excited (fluorescent) state.[17] The following correlation function for the incident photon field is used:

$$\langle E^{(+)}(t_1)E^{(-)}(t_2)\rangle = \overline{\epsilon}(t_1)\overline{\epsilon}(t_2)\exp[i\omega_I(t_1-t_2)]\widetilde{G}(t_1-t_2) \ , \tag{3.35}$$

where the stationary function $\widetilde{G}$ is assumed to take the form

$$\widetilde{G}(t_1-t_2) = \exp[-|t_1-t_2|\Gamma'] \ , \tag{3.36}$$

in which $1/\Gamma'$ represents the correlation time of the photon source, and $\overline{\epsilon}(t)$ is the pulse amplitude and is assumed to be given by

$$\overline{\epsilon}(t) = \exp(\gamma_1 t/2) \ , \qquad\qquad t < 0 \ ,$$

$$= 1 \ , \qquad\qquad 0 \le t \le \tau_p \ , \tag{3.37}$$

$$= \exp[-\gamma_2(t-\tau_p)/2] \ , \qquad \tau_p < t \ .$$

In Figs. 3.8a and 3.8b, the fluorescence intensity denoted by solid line is calculated together with the degree of the molecular coherence between the two excited states (broken line), $C(m,n)$ defined by Eq. (2.62). In these figures, the effect of the pure dephasing constant between the excited states, $\Gamma_{mn}^{(d)}$ is shown. In Figs. 3.8a and 3.8b, $\Gamma_{mn}^{(d)} = 0$ and $\Gamma_{mn}^{(d)} = 0.2\omega_{mn}$ were used. Other constants used were $\Gamma' = 0$, $\tau_p = 3.14/\omega_{mn}$, $\Gamma_{ma}^{(d)} = \Gamma_{na}^{(d)} = 0.1\omega_{mn}$, $\Gamma_{mm} = \Gamma_{nn} = 0.1\omega_{mn}$, $\Gamma_{aa} = \Gamma_{ff} = 0$, $M_{ma} = M_{na} = M_{mf} = M_{nf}$, $\gamma_1 = \gamma_2 = 10.0\omega_{mn}$, and $\omega_I - \omega_n = 0.5\omega_{mn}$ as an excitation condition. The decay of the quantum beat after the pulse duration in Fig. 3.8a is due to the population decay constants Γ_{mm} and Γ_{nn}. From both figures, we can see that the electronic pure dephasing constant $\Gamma_{ma}^{(d)}$ does not attenuate C(m,n) after the pulse duration. The vibrational pure dephasing constant between the excited states, on the other hand, attenuates C(m,n) during and after the pulse duration as well as the intensity of the quantum beat. Effects of the photon field coherence have also been investigated;[17] Γ' shows the same dependence on the fluorescence intensity as that of the electronic dephasing between the initial and excited states, i.e., during the pulse duration Γ' operates.

III-3. Time-resolved resonance Raman scattering from molecules with a nonequilibrium vibronic distribution

In this section, a theoretical treatment of the time-resolved resonance Raman scattering(RRS) from vibronic levels of a molecule in a nonequilibrium condition is presented.[51,52] Time-resolved resonance Raman spectroscopy is now widely applied to studying dynamics of short-lived vibronically excited molecules such as photochemical and biological intermediates and so on.[53-56] The spectroscopy basically consists of the measurement of the time-dependent population of the vibronically excited states by controlling the time delay between excitation laser and the probe Raman scattering. The vibrational spectra of the time-resolved RRS depend on the initial nonequilibrium vibronic distribution which is produced by optical excitations, electronic relaxations, photochemical reactions following the optical excitation, etc., and on its relaxation mechanism which governs the subsequent population change.

Theoretical treatments of the time-resolved RRS from nonequilibrium species have been reported by Luzzi and Vasconcellos,[57] and Vasconcellos and Luzzi.[58] They have developed the theory of the time-resolved Raman scattering from a highly photo-excited semi-conductor plasma by using the Zubarev nonequilibrium ensemble method. In this section, the generating function method [59,60] is applied to deriving an expression for the RRS differential cross section.

Let us consider a time-resolved RRS from initial vibronic states $a\{v\}$ to the final states $a\{v'\}$ via resonant electronic states. In the case in which the pumping and probe lasers are independent each other, that is, coherent effects can be neglected, and the probe Raman laser does not change the population, the differential cross section in a solid angle $d\Omega$ per unit scattered photon frequency ω_2, $d^2\sigma(\omega_1,\omega_2,t)/d\Omega d\omega_2$ is defined as

$$\frac{d^2\sigma(\omega_1,\omega_2,t)}{d\Omega d\omega_2} = \sum_v \rho_v(t)\frac{d^2\sigma_v(\omega_1,\omega_2)}{d\Omega d\omega_2} \quad , \tag{3.38}$$

where $\rho_v(t)$ represents the time-dependent vibronic distribution of the initial electronic state and is determined by solving the master equation. In Eq.(3.38), ω_1 is the incident photon frequency, and $d^2\sigma_v(\omega_1,\omega_2)/d\Omega d\omega_2$ is the RRS differential cross section from single vibronic level (SVL) v. The differential cross section from the SVL is given by

$$\frac{d^2\sigma_v(\omega_1,\omega_2)}{d\Omega d\omega_2} = C\omega_1\omega_2^3 \sum_{v'} \frac{\Gamma_{av',av}}{(\omega_{av',av} - \omega_1 +\omega_2)^2+\Gamma^2_{av',av}}$$

$$\times | \sum_m \frac{M_{av',bm}(2)M_{bm,av}(1)}{i(\omega_{bm,av}-\omega_1)+\Gamma_{bm,av}} |^2 \quad , \tag{3.39}$$

where $C=(2\pi\hbar^2\varepsilon_0 c^4)^{-1}$. In the Born-oppenheimer and the Condon approximations, Eq.(3.39) can be rewritten in terms of the generating function as

$$\frac{d^2\sigma_v(\omega_1,\omega_2)}{d\Omega d\omega_2} = \frac{C}{2} \omega_1\omega_2^3 |M_{ab}^0(2)M_{ba}^0(1)|^2 \int_0^\infty d\tau \int_0^\infty d\tau' \int_{-\infty}^\infty d\xi$$

$$\times \exp[i\xi(\omega_1-\omega_2)+i\omega_1(\tau'-\tau)-\Gamma_{ba}(\tau+\tau')-\Gamma_{v'v}|\xi|]G_v(\tau,\tau',\xi),$$

$$\tag{3.40}$$

where the generating function for the single vibronic level
RRS, $G(\tau,\tau',\xi)$ is expressed as

$$G_v(\tau,\tau',\xi) = \langle v|\exp[i\hat{H}_b\tau/\hbar]\exp[-i\hat{H}_a\xi/\hbar]\exp[-i\hat{H}_b\tau'/\hbar]$$

$$\times \exp[i\hat{H}_a(\xi-\tau+\tau')/\hbar]|v\rangle. \tag{3.41}$$

In Eq.(3.41), the vibrational quantum number dependence of
the dephasing constants has been omitted, that is Γ_{ba} and
$\Gamma_{v'v}$ are the average electronic and vibrational dephasing
constants, respectively. $\hat{H}_a$ and $\hat{H}_b$ in Eq.(3.41) represent
the vibrational Hamiltonians of the electronic states, a and
b, respectively. For simplicity, we restrict ourselves to a
model Hamiltonian for a displaced harmonic oscillator with
frequency ω, dimensionless displacement Δ, and electronic energy
gap ϵ^0. In this model, the generating function for the single
vibronic RRS is given by[51]

$$G_v(\tau,\tau',\xi) = \exp[i\epsilon^0(\tau-\tau')/\hbar + \frac{\Delta^2}{2}\{\exp(i\omega\tau)-1\}$$

$$+ \frac{\Delta^2}{2}\{\exp(-i\omega\tau')-1\} + \lambda^*(\tau)\lambda(\tau')\exp(-i\omega\xi)]L_v(|\alpha(\tau,\tau',\xi)|^2),$$
$$\tag{3.42}$$

in which L_v is the Laguerre polynomials,

$$\alpha(\tau,\tau',\xi) = \lambda(\tau) - \lambda(\tau')\exp(-i\omega\xi) \quad , \tag{3.43}$$

and

$$\lambda(\tau) = \frac{\Delta}{\sqrt{2}} \{1 - \exp(-i\omega\tau)\} \quad . \tag{3.44}$$

Substituting Eq.(3.40) with Eq.(3.42) into Eq.(3.38) yields

$$\frac{d^2\sigma(\omega_1,\omega_2,t)}{d\Omega d\omega_2} = -\frac{C}{2}\, \omega_1\omega_2^3\, |M_{ab}^0(2)M_{ba}^0(1)|^2 \int_0^\infty d\tau \int_0^\infty d\tau' \int_{-\infty}^\infty d\xi$$

$$\times \exp[i\xi(\omega_1-\omega_2)+i\omega_1(\tau'-\tau)-\Gamma_{ba}(\tau+\tau')-\Gamma_{v'v}|\xi|]G(\tau,\tau',\xi;t),$$
$$\tag{3.45}$$

where $G(\tau,\tau',\xi;t)$ is the generating function for the time-resolved RRS, and is defined as

$$G(\tau,\tau',\xi;t) = \sum_v \rho_v(t)G_v(\tau,\tau',\xi) \quad . \tag{3.46}$$

Let us consider a time-resolved RRS from molecules in which the vibronic distribution is prepared only at level v_0 at t=0 (single vibronic level distribution, and relaxes by undergoing both electronic and collisional vibrational relaxations:

$$M_{v_0} \xrightarrow{\quad k_v^c \quad} M_{v_0-1} \xrightarrow{\hspace{2cm}} \ ; \quad M_v \xrightarrow{\quad k_v^e \quad} P \ .$$

If these relaxation constants are assumed to be a linear function of the vibrational quantum number v, that is, $k_v^c = vk_c$ and $k_v^e = vk_e$, the single vibronic distribution at time $t, \rho_v(t)$ is given by[61]

$$\rho_v(t) = \rho_{v_0}(0) \frac{v_0!}{v!(v_0-v)!} [-\frac{k_c}{k_c+k_e}\{1-\exp[-(k_c+k_e)t]\}]^{v_0-v}$$

$$\times \ \exp[-v(k_c+k_e)t] \ , \tag{3.47}$$

where $\rho_{v_0}(0)$ is the density of the v_0 vibronic state at t=0. Substituting Eq.(3.47) into Eq.(3.46) and taking summation over v in the resulting expression, we can obtain the expression for the generating function. Substitution of the expression for the generating function into Eq.(3.45) yields the final expression for the time-resolved RRS from the SVL excited molecules as

$$\frac{d^2\sigma(\omega_1,\omega_2,t)}{d\Omega d\omega_2} = C\omega_1\omega_2^3|M_{ab}^0(2)M_{ba}^0(1)|^2\rho_{v_0}(0) \sum_{j=0}^{\infty} \sum_{n=0}^{v_0} \sum_{p=0}^{n} \sum_{q=0}^{n}$$

$$\times \frac{(-1)^{p+q}}{j!\ n!}\ \binom{v_0}{n}\binom{n}{p}\binom{n}{q}\{\frac{k_c}{k_c+k_e}+\frac{k_e}{k_c+k_e}\exp[-(k_c+k_e)t]\}^{v_0-n}$$

$$\times \exp[-n(k_c+k_e)t]\ \frac{\Gamma_{v'v}}{[\{\omega_1-\omega_2+\omega(n-p-q-j)\}^2+\Gamma^2_{v'v}]}$$

$$\times F(\omega_1)^{(p+j,n-q)}\ F(\omega_1)^{*(q+j,n-p)}\ ,\tag{3.48}$$

where

$$F(\omega_1)^{(k,\ell)}=\int_0^\infty d\tau\{\lambda^*(\tau)\}^k\{\lambda(\tau)\}^\ell\ \exp[i(\varepsilon^0/\hbar-\omega_1)\tau+\frac{\Delta^2}{2}\{e^{i\omega\tau}-1\}-\Gamma_{ba}\tau].$$

$$\tag{3.49}$$

This function can be evaluated by using a numerical integration or analytically: the analytical expression is given by

$$F(\omega_1)^{(k,\ell)}=\exp(-\frac{\Delta^2}{2})(\frac{\Delta}{\sqrt{2}})^{k+\ell}\sum_{n=0}^\infty\sum_{r=0}^k\sum_{u=0}^\ell\frac{(-1)^{r+u}}{n!}$$

$$\times(\frac{\Delta^2}{2})^n\binom{k}{r}\binom{\ell}{u}\frac{1}{i\{\omega_1-\varepsilon^0/\hbar+(u-r-n)\omega\}+\Gamma_{ma}}\ .\tag{3.50}$$

To demonstrate the time-dependence of the vibrational intensity of the RRS in the case of the single vibrational level distribution as the initial distribution, a model calculation of the Raman spectra by using Eq.(3.46) has been performed. Figure 3.9 shows the calculated Raman spectra of the model molecule initially excited to $v_0=3$ level with potential displacement $\Delta=1.0$, vibrational frequency $\omega=100$ cm^{-1}. $\varepsilon^0/\hbar -\omega_1 =300$cm^{-1} is taken as the excitation condition. The ordinate I denotes the Raman intensity relative to the maximum band intensity without the optical frequency factors $\omega_1\omega_2^3$, and $\Gamma_{ma}=100$cm^{-1} and $\Gamma_{v'v}=50$cm^{-1} are taken as the dephasing constants. This figure represents the calculated Raman spectra in which the contribution of the electronic relaxation can be neglected compared with that of the collisional vibrational relaxation, that is , $k_e=0$. the solid, dotted-broken and broken lines denote the Raman spectra at $k_c t=0$, 0.2, and 0.4, respectively. It should be noted that the first through third order anti-Stokes Raman bands make a significant contribution to the spectra. The appearance of the anti-Stokes bands is mainly related to which vibronic level is initially prepared. We can see that because of the vibrational relaxation effect the Raman intensities of both anti-Stokes and Stokes bands decrease, and ,on the other hand, the Rayleigh band intensity increases as $k_c t$ increases.

So far time-resolved RRS from a SVL system has been presented. Analysis for the time-resolved RRS from molecules characterized by other initial vibronic distributions, such as Poisson and canonical distribution have been reported.[51,52]

A model calculation of the excitation profiles of the time-resolved RRS from molecules with nonequilibrium distributions has been given in Ref.52. The results of the calculation indicate that the anti-Stokes Raman bands make a significant contribution to the resonance Raman scattering from the nonequilibrium vibronic distributions. It is suggested that measurement of the excitation profiles are useful for analyzing the nonequilibrium distribution initially prepared and the subsequent relaxation mechanism.

In summary, in this chapter,a kinetic equation for the multiphoton transition of molecules and the expressions for the transition probabilities have been derived based on the master equation approach in which dephasing effects are considered together with the molecule-photon interaction. A two-photon transition probability via coherently excited vibronic levels has been formulated.Mechanisms of the multiphoton transition have been investigated; sequential multiphoton transitions induced by the electronic dephasings are predicted for the resonant two- and three-photon absorptions. An expression for the vibronic intensity distribution for the two-photon transition, which depends on its mechanism, has been evaluated within the Born-Oppenheimer approximation. Effects of the vibrational relaxations for a resonant two-photon transition, and dephasing effects in quantum beats in molecular fluorescence have been investigated.

It is shown that the time-resolved RRS may play an important role in studying relaxation mechanisms of the molecule with a nonequilibrium distribution.

Acknowledgments

The author is indebted to Professor S. H. Lin for many useful comments and interesting discussions, to Professor T. Nakajima for his encouragement and helpful suggestions, and to Dr. H. Kono for stimulating discussions.

References

1. D. H. Parker, J. O. Berg, and M. A. El-Sayed, Springer Ser. Chem. Phys. 3, 320 (1978).

2. P. M. Johnson, Acc. Chem. Res. 13, 20 (1980).

3. P. M. Johnson and C. E. Ottis, Annu. Rev. Phys. Chem. 32, 139 (1981).

4. R. B. Bernstein, J. Phys. Chem. 86, 1178 (1982).

5. E. W. Schlag and H. J. Neusser, Acc. Chem. Res. 16, 355 (1983).

6. K. Kimura, "Photoelectron spectroscopy of excited states" in Adv. in Chem. Phys. 1984.

7. D. A. Gobeli, J. J. Yang and M. A. El-Sayed, "Some studies on laser multiphoton ionization and multiphoton ionization dissociation of polyatomic molecules" in "Adv. in multi-photon processes and spectroscopy" ed. by S. H. Lin (World Scientific, Singapore, 1984) p.51.

8. I. Tanaka and M. Kawasaki, " ultiphoton processes by visible and UV lasers" ibid. p.239.

9. S. H. Lin, Y. Fujimura, H. J. Neusser and E. W. Schlag, "Multiphoton spectroscopy of molecules" Academic Press, Orlando (1984).

10. S. Nakajima, Prog. Theor. Phys. 20, 948 (1958).

11. R. Zwanzig, Physica (Utrecht) 30, 1109 (1964).

12. S. H. Lin and H. Eyring, Proc. Natl. Acad. Sci. U.S.A. 74, 3105 (1977); ibid. 74, 3623 (1977).

13. R. Kubo in Fluctuation, rexation, and resonance in

magnetic systems, Ed. by D. ter Haar (Oliver and Boyd, London, 1962) p.23.

14. R. Kubo, J. Phys. Soc. Japan 17, 1100 (1962).

15. J. H. Freed, J. Chem. Phys. 49, 376 (1968).

16. S. Mukamel, I. Oppenheim, and J. Ross, Phys. Rev. A17, 1988 (1977).

17. H. Kono, Y. Fujimura and S. H. Lin, J. Chem. Phys. 75, 2569 (1981).

18. K. E. Jones and A. H. Zewail, in Adv. in laser Chemistry, ed. by A. H. Zewail (Springer series in Chemical Physics, Springer-Verlag, New York, 1978), p.196.

19. T. Takagahara, E. Hanamura, and R. Kubo, J. Phys. Soc. Japan 43, 802 (1977).

20. T. Takagahara, E. Hanamura, and R. Kubo, J. Phys. Soc. Japan, 44, 728 (1978).

21. E. Hanamura and T. Takagahara, J. Phys. Soc. Japan 47, 410 (1979).

22. Y. Fujimura and S. H. Lin, J. Chem. Phys. 75, 5110 (1981).

23. J. Murakami, K. Kaya, and M. Ito, J. Chem. Phys. 72, 3263 (1980).

24. Y. Fujimura and S. H. Lin, J. Chem. Phys. 74, 3726 (1981).

25. D. Lee and A. C. Albrecht, "A unified view of Raman scattering, resonance Raman scattering and resonance fluorescence (and their analogues in two-photon absorption)" in Adv. in Infrared and Raman spectroscopy ed. by R. J. H. Clark and R. E. Hester (Heyden, London,

1985).

26. V. Hizhnyakov and I. Tehver, Phys. Status Solidi 21, 755 (1967).

27. D. L. Huber, Phys. Rev. 158, 843 (1967).

28. Y. R. Shen, Phys. Rev. B9, 622 (1974).

29. S. Mukamel, A. Ben-Reuven, and J. Jortner, J. Chem. Phys. 64, 3971 (1976).

30. R. M. Hochstrasser and F. A. Novak, Chem. Phys. Lett. 48, 1 (1977); 53, 3 (1978).

31. Y. Fujimura, H. Kono, T. Nakajima and S. H. Lin, J. Chem. Phys. 75, 99 (1981).

32. H. Kono, Y. Fujimura and T. Nakajima, Int. J. Quant. Chem. 18, 293 (1980).

33. Y. Fujimura and S. H. Lin, J. Chem. Phys. 78, 6468 (1983).

34. K. K. Lehman, J. Smolarek, and L. Goodman, J. Chem. Phys. 69, 1569 (1978).

35. M. Kawasaki, K. Tsukiyama, M. Kuwana, K. Obi, and I. Tanaka, Chem. Phys. Lett. 67, 365 (1979).

36. K. Kasatani, ·Y. Tanaka, K. Shibuya, M. Kawasaki, K. Obi, H. Sato, and I. Tanaka, J. Chem. Phys. 74, 895 (1981).

37. P. Cremaschi, Chem. Phys. Lett. 83, 106 (1981).

38. P. M. Johnson, J. Chem. Phys. 64, 4638 (1976).

39. P. M. Johnson, J. Chem. Phys. 64, 4143 (1976).

40. K. Sato, Y. Achiba, and K. Kimura, J. Chem. Phys. 81, 57 (1984).

41. J. Chaiken, M. Gurnick, and J. D. McDonald, Chem. Phys.

Lett. 61, 197 (1979); J. Chem. Phys. 74, 106 (1981).

42. H. Henke, H. L. Selzle, T. R. Hays, S. H. Lin, and E. W. Schlag, Chem. Phys. Lett. 77, 448 (1981).

43. P. M. Felker, W. R. Lambert, and A. H. Zewail, Chem. Phys. Lett. 89, 309 (1982); W. R. Lambert, P. M. Felker, and A. H. Zewail, J. Chem. Phys. 81, 2217 (1984).

44. W. Shafin, M. Ivanco, and S. C. Wallace, J. Chem. Phys. 76, 2095 (1982); M. Ivanco, J. Hager, W. Sharfin, and S. C. Wallace, J. Chem. Phys. 78, 6531 (1983).

45. S. Okajima, H. Saigusa and E. C. Lim, J. Chem. Phys. 76, 2096 (1982).

46. B. J. van der Meer. H. Th, Jonkman, G. M. ter Horst, and J. Kommandeur, J. Chem. Phys. 76, 2099 (1982).

47. H. Watanabe, S. Tsuchiya, and S. Koda, J. Phys. Chem. 87, 906 (1983).

48. P. J. Brucat and R. N. Zare, J. Chem. Phys. 78, 100 (1983).

49. E. W. Schlag, E. W. Henke, and S. H. Lin, Int. Rev. Phys. Chem. 2, 43 (1982).

50. Y. Fujimura, T. Akiyama, T. Nakajima, H. Kono, and S. H. Lin (To be published).

51. Y. Fujimura, Y. Ohtsuki and T. Nakajima, Bull. Chem. Soc. Jpn. 1985 in press.

52. Y. Fujimura, Y. Ohtsuki, M. Arai, and S.H.Lin ,J.Chem. Phys.1985,in press.

53. G. H. Atkinson, in Adv. in Infrared and Raman Spectry, ed. by R. J. H. Clark and R. E. Hester (Heyden,London, 1982), 9, p.1.

54. T.Kobayashi and K. Yokoyama, Kagaku No Ryoiki 36,509(1982).

55. G. H. Atkinson, Time-resolved Vibrational Spectroscopy (Academic, New York,1983).

56. Y. Fujimura, Y. Ohtsuki, and T. Nakajima, Proc. Ninth Int. Conf. Raman Spectry(Tokyo,1984) p678.

57. R.Luzzi and A. R. Vasconcellos, J. Raman Spectry.10,28 (1981).

58. A. R. Vasconcellos and R. Luzzi,ibid.,14,39(1983).

59. Y.Fujimura and S. H. Lin, J.Chem. Phys. 70,247(1979).

60. Y.Fujimura and S. H. Lin,ibid,71,3733(1979).

61. Y.Fujimura, S. H. Lin, H. Schröder, H. J. Neusser,and E.W.Schlag, Chem.Phys. 43,205(1979).

Figure captions

Fig.2.1.   Diagramatic representations of the time evolution of ket and bra vectors related to the one-photon absorption. The diagram in the left-hand side and that in the right-hand side correspond to the first and second terms in Eq.(2.61), respectively.

Fig.2.2.   Diagramatic representations of the ket and bra vectors for a two-photon absorption via coherenty excited vibronic levels. Figures 2.2a ,2.2b, and 2.2c correspond to the time evolution in Eqs.(2.69a),(2.69b),and(2.69c), respectively.

Fig.3.2.   A simultaneous two-photon transition of a molecule in the displaced harmonic potential model.

Fig.3.3   The calculated vibronic intensity distribution for the two-photon absorption as a function of $\omega_2 - (\epsilon_f^0 - \epsilon_m^0)/\hbar$ in the case of $\Delta_{fm} = 2.0$. The doublet structure reflects the simultaneous and sequential mechanisms; the bands whose peak locates at the integral multiple on the abscissa correspond to those originated from the sequential mechanism, and the other bands correspond to those originated from the simultaneous mechanism.

Fig.3.4.   The calculated vibronic intensity distribution for the two-photon absorption in the case of $\Delta_{fm} = 0$.

Fig.3.5   Resonant (2+1) and (1+2) three-photon absorption processes. {m} and {k} represent the resonant intermediate and the virtual states, respectively.

Fig.3.6.   The calculated vibronic intensity distribution for a resonant (2 + 1) absorption process. The vibronic bands denoted by a and b mainly originate from the rate constants, $k^{(2+1,sim)}$ and $k^{(2+1,seq)}$, respectively.

Fig.3.7.   The mechanism of the sequential absorption induced by the electronic dephasing process related to Fig.3.6.

Fig.3.8a.   The effect of the pure electronic dephasings $\Gamma_{ma}^{(d)}$ and $\Gamma_{na}^{(d)}$ on fluorescence intensity $I(t)$ and the degree of molecular coherence $C(m,n)$. The solid and broken lines represent $I(t)$, and $C(m,n)$, respectively.

Fig.3.8b.   The effect of the pure dephasing $\Gamma_{mn}^{(d)}$ on $I(t)$, and $C(m,n)$.

Fig.3.9.   The calculated time-resolved Raman spectra of the model molecule with single vibronic distribution $v_0=3$. The solid, dotted-broken, and broken lines denote the Raman spectra at $k_c t=0$, 0.2, and 0.4, respectively.

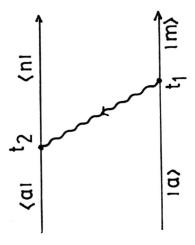

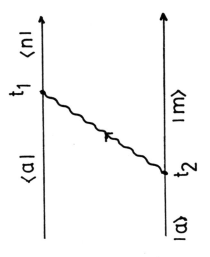

Fig. 2.1

68

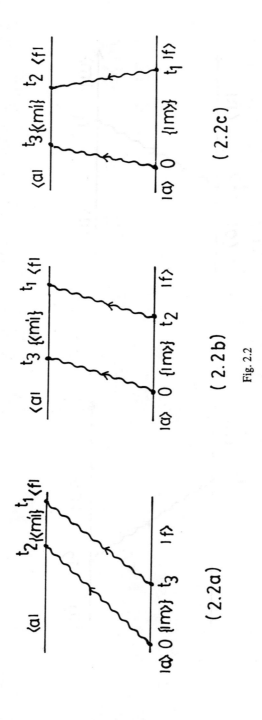

$(2.2a)$

$(2.2b)$

$(2.2c)$

Fig. 2.2

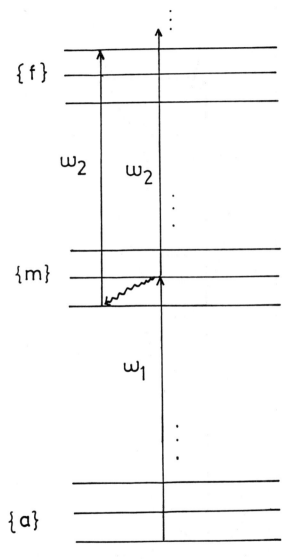

Fig. 3.1

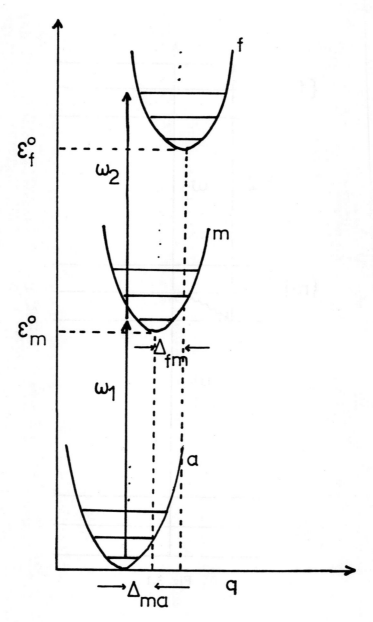

Fig. 3.2

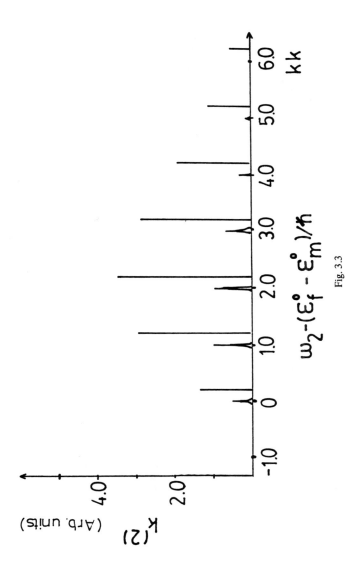

Fig. 3.3

72

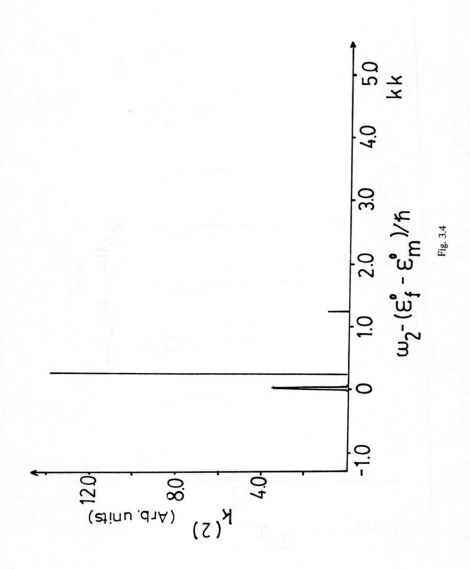

Fig. 3.4

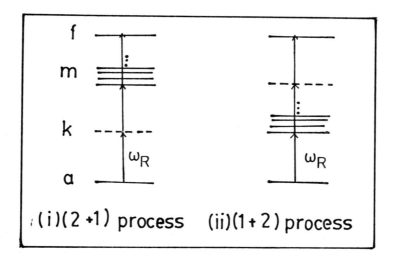

Fig. 3.5

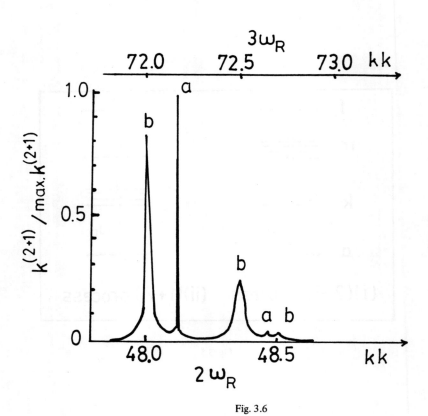

Fig. 3.6

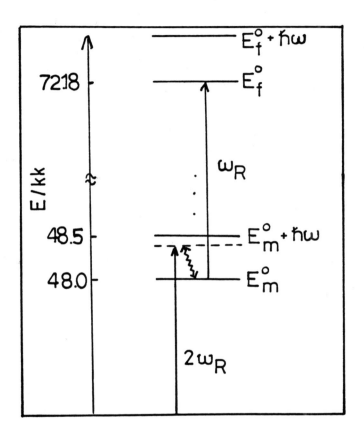

Fig. 3.7

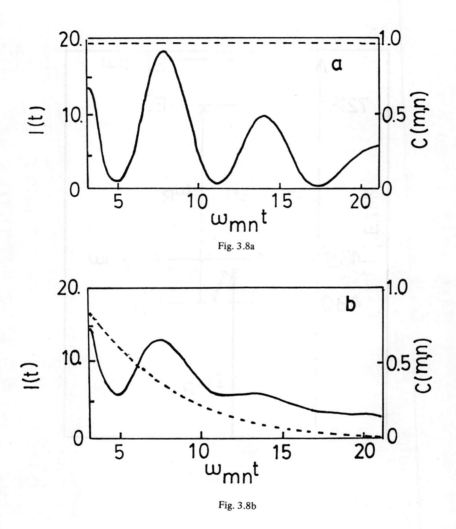

Fig. 3.8a

Fig. 3.8b

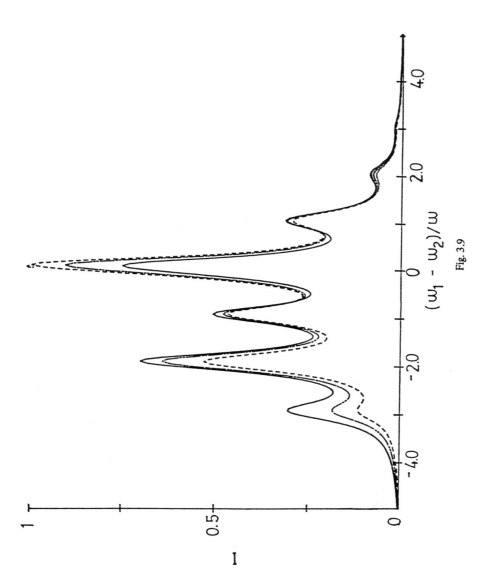

Fig. 3.9

2. PHOTOCHEMISTRY, PHOTOPHYSICS, AND SPECTROSCOPY OF MOLECULAR INFRARED MULTIPLE PHOTON EXCITATION

J. S. Francisco* and J. I. Steinfeld

Department of Chemistry
Massachusetts Institute of Technology
Cambridge, Massachusetts 02139, USA

We review infrared multiple-photon absorption, excitation, and dissociation in small polyatomic molecules, with an emphasis on the chemical consequences of these processes. A survey of theoretical models for IRMPE is presented, touching upon the fundamental question of intramolecular vibrational relaxation which is raised by these models. What experimental probes are available for following the dynamics of IRMPE? Bulk-averaged measurements, while convenient, provide only limited information; direct spectroscopic probes of time-dependent level populations have been attempted, but may be difficult to interpret for levels higher than the first few vibrational fundamentals. Finally, what useful chemical knowledge and/or products have resulted from IRMPE? Significant chemical-kinetic information has resulted from using this method to initiate reactions; however, its synthetic utility has been extremely limited until the recent interest in laser-driven surface reactions, which are beginning to find important applications in processing of electronic materials.

*Present address: Department of Chemistry, Wayne State University, Detroit, Michigan 48202, USA.

Contents

1. INTRODUCTION

A. Unimolecular Reactions and the Radiation Hypothesis

The rate at which chemical transformations occur depends on a number of parameters--species concentrations, presence of inert diluents (buffer gases or solvents), temperature, pressure, etc. A common feature is the requirement for external energy input, in order to overcome an activation barrier intervening between reactants and products. This input can be an increase in the energy distributed throughout the system at thermodynamic near-equilibrium (i.e., raising the temperature), or in the form of a specific energy input such as relative kinetic energy (in accelerated molecular beams), ionizing radiation, or light. The photochemical initiation of reactions was limited to u.v. atomic lamps until only recently, when powerful laser systems made it possible to deliver appreciable amounts of energy to reacting systems at a variety of u.v., visible, and even infrared wavelengths.

By the beginning of the present century, the phenomenological behavior of chemical reaction rates was fairly well established. The dependence on temperature was given by the Arrhenius equation,

$$k_{react} = A \exp(-E_{act}/RT) \tag{1.1}$$

A variety of dependences on species concentrations, or reaction orders, had been established. One large and important class of gas-phase reactions had been found to follow first-order kinetics, that is, the disappearance rate of a particular component of the system was given by

$$\frac{-d[A]}{dt} = k_{uni}[A]. \tag{1.2}$$

These reactions, typically dissociation or isomerization of species A, were assigned to a unimolecular mechanism, that is, one in which the rate-determining step involved just one molecule of A. Since the activation

energies for these reactions are substantial - typically on the order of

100-400 kJ/mole - the question naturally arose as to how such a molecule was

able to acquire the energy required to undergo reaction.

One of the first suggestions was put forward by Perrin [1919].

Unimolecular reactions are typically carried out in heated vessels enclosing

the reacting gaseous system. Perrin proposed that black-body radiation from

the hot walls of the vessel was absorbed by the molecules, causing them to

react. This hypothesis was rapidly abandoned; many physical chemistry

textbooks claim that Langmuir [1920] proved that the amount of energy radiated

would be insufficient to account for the observed reaction rates. In fact,

the argument advanced by Langmuir was a specious one. He considered only

frequencies of radiation with $h\nu = hc/\lambda > E_{act}$. For $E_{act} \sim 300$ kJ/mole, this

corresponds to $\lambda < 400$ nm, and the blackbody amplitude is essentially zero at

these wavelengths, even at $T \sim 1000K$. Most molecules, however, possess strong

absorption bands in the infrared region, i.e., between 1 and 10 μm, and the

peak of the blackbody curve lies in just this wavelength region for typical

temperatures (300K to 2000K). A simple calculation using the Wien radiation

law

$$E_\nu d\nu = \frac{2\pi h\nu^3}{c^2} e^{-h\nu/kT} d\nu \tag{1.3}$$

shows that more than enough infrared energy is radiated into the gas to

account for the observed rates. Of course, the molecule would have to absorb

several photons at these wavelengths in order to accumulate sufficient energy

to react.

Why then should one have abandoned Perrin's radiation hypothesis? The

real reason is a thermodynamic one, which was completely neglected by

Langmuir. The radiation mechanism requires that one system (a blackbody at T) do chemical work on another system (reacting gas) at the same temperature T. The second law states that this is impossible without a reservoir at some lower T to absorb heat. Physically, if the gas and the heated wall are at the same temperature, the gas will radiate just as much energy back to the wall as it absorbs, and thus no net reaction is possible. A source at a much higher temperature is necessary to effect dissociation; until lasers became available, such sources did not exist.

There are other clear objections as well to the Perrin radiation hypothesis. Langmuir's argument does apply to a number of molecules, such as the diatomic halogens, which undergo unimolecular dissociation but do not possess infrared absorption bands. Most importantly, the radiation hypothesis predicts k_{uni} to be truly independent of pressure, but it is well established that there is a low-pressure falloff of the rate coefficient for most reactions. For all these reasons, the Lindemann-Hinshelwood collisional activation mechanism was adopted, and Perrin's radiation hypothesis was largely forgotten.

If it were possible to separate the system into a high-temperature radiation source and a cool reacting gas, then radiation activation would be not just feasible, but a very efficient way of initiating unimolecular reactions. The laser, which can be regarded as a source with an extremely high equivalent radiation temperature, has made this possible. The key is a luminous intensity sufficiently high to induce multiphoton absorption in the molecule; in this way, a large number of low-energy photons can contribute to a high-energy process such as breaking a chemical bond. The first reaction identified as a multiphoton dissociation actually used visible, rather than infrared light; the Q-switched output pulse of a ruby laser was found to decompose phthalocyanine vapor, while the free-running pulse with higher total energy but lower peak power was ineffective [Porter and Steinfeld, 1966]. The first use of an infrared laser to drive a unimolecular reaction in SiF_4 was reported by Isenor and co-workers [1971, 1973]; what really attracted attention, and established infrared multiple-photon dissociation (IRMPD) as an important process, was the isotopically selective dissociation of SF_6 reported by Soviet and U.S. groups in 1975 [Ambartzumian et al., 1975; Lyman et al., 1975]. In the past ten years, a wide range of molecules has been found to undergo IRMPD [Steinfeld, 1981].

B. Infrared Multiple-Photon Absorption, Excitation, and Dissociation

The process described above has been alluded to by a variety of terms in the literature. In this review, we shall employ the terms Infrared Multiple-Photon Absorption (IRMPA) to refer to the interaction of a molecule with intense infrared radiation; Infrared Multiple-Photon Excitation (IRMPE) to denote the resulting excitation of the molecule; and Infrared Multiple-Photon Dissociation (IRMPD) if the excitation is sufficiently great to exceed a dissociation threshold. In the pioneering work by Isenor and co-workers [Isenor et al., 1971, 1973], these processes were detected as visible fluorescence from gaseous samples of SiF_4 irradiated by a high-intensity CO_2 laser. The luminescence resembled the infrared laser pulse in temporal and geometric shape, and the authors were able to assign the fluorescence to excited photofragments. These observations, coupled with the additional facts that: 1) there are no fundamental vibrational frequencies of SiF_4 which coincide with the CO_2 laser frequency used for excitation and 2) the laser pulses were 200 ns in duration, which is less than the average time between gas kinetic collisions in the low pressure samples, led the authors to present an interpretation which stated that sufficient energy was absorbed by near resonant vibrational modes of the molecule to cause direct collisionless dissociation. Furthermore, absorption of the requisite number of IR photons was suggested to occur through a series of discrete vibrational levels. As the vibrational energy content of the molecule increases, so does the density of vibrational states (number of states per unit vibrational energy). This, combined with anharmonic coupling and Coriolis splitting of degenerate vibrational levels causes the vibrational states to become very dense, a situation described by the term "quasi-continuum" (QC). It is the quasi-continuum, as the authors suggested, that provides a facile pathway for absorption of energy by the molecule from the laser field, since resonance with the laser frequency is assured by the high density of states. This study not only provided the first postulate of the quasi-continuum, but also explained its role in allowing IRMPA to overcome bottlenecks which may arise as a result of anharmonicity.

Since this work, an intensive study of the mechanism by which a molecule could absorb multiple infrared photons and dissociate has been carried out by many groups. Although research was stimulated largely by the potential application of the process to laser isotope separation [Cantrell, 1979; Letokhov and Moore, 1977] the intense efforts did lead to an understanding of the mechanism of IRMPA and IRMPD for polyatomic molecules. There are several reviews which give a useful chronology of experiments and interpretation of the IRMPA and IRMPD process [Ambartzumian and Letokhov, 1977a,b; Wolfrum, 1977; Kimel and Speiser, 1977; Bauer, 1978; Grunwald et al., 1978; Kompa et al., 1978; Schulz et al., 1979; Quack, 1982].

The question which shaped earlier views of the multiphoton absorption process was, how could a molecule absorb additional photons from the laser field after the absorption of the first photon, since the anharmonicity of the vibrational mode would take the molecule off-resonance. Bloembergen [1975] provided the seminal suggestion that near resonant fields could strongly excite an off-resonant pump mode in a coherent fashion with laser intensities used in typical IRMPA experiments. From experiments which used sensitive probes of the spectroscopy and molecular dynamics of IRMPE, such as the two-frequency pump-probe discussed in Section 3B and two-frequency multiphoton excitation and dissociation experiments [Ambartzumian et al., 1977; Jensen et al., 1979; Gower and Gustafson, 1977b; Moulton et al., 1977], along with several modified theoretical models [Larsen and Bloembergen, 1976; Larsen, 1976; Cantrell and Galbraith, 1977; Ackerhalt and Galbraith, 1978; Quack, 1978, 1982], a mechanism emerged for the excitation process through the low energy molecular vibrational manifold (Region I as shown in Figure 1.1). The process is characterized by coherent excitation for near resonant laser fields through a set of discrete levels. The difficulty of continued excitation posed by bottlenecking due to anharmonicity is compensated by several processes such as rotational level compensation through power broadening, anharmonic splitting of multiple-degenerate modes, and Coriolis splitting in excited rovibrational states; which then allow facile excitation into the quasicontinuum.

In order to connect the discrete and QC regions into a unified IRMPE picture, several researchers have explored the transition between the two regions with particular reference to the onset of the quasicontinuum [Quack, 1978; Ackerhalt and Galbraith, 1978]. For a given molecule, it has been shown that the QC onset is determined by the condition

$$\Delta\omega_{mol} \gg \Omega_L \gg \rho^{-1}(E) \tag{1.4}$$

where $\Delta\omega_{mol}$ is the molecular linewidth, Ω_L is the Rabi frequency $\hbar^{-1}\underline{\mu}\cdot\underline{E}$, and

$\rho(E)$ defines the density of vibrational levels at a specific energy. Several models are now available which bridge the discrete and quasicontinuum regions right through to dissociation [Mukamel and Jortner, 1976b; Stone et al., 1976; Stone et al., 1979; Horsley et al., 1979; Galbraith and Ackerhalt, 1981], but the Galbraith and Ackerhalt model provides the closest connection with a true spectroscopic model.

The general features of all these models can be summarized as follows. Once in the quasicontinuum, defined as Region II in Figure 1.1, the interaction between the molecule and the laser field ceases to be coherent, and the dynamics of the excitation process can be described by a set of coupled linear rate equations for each state, i.e., by a Master Equation [Oppenheim et al., 1977]. If each vibrational state were to be included, the resulting number of equations would render the Master Equation unsolvable (e.g., at 60 kJ mol^{-1} internal excitation, a typical molecule may have somewhere between $10^3 \sim 10^4$ states per cm^{-1}); however, the problem can be substantially reduced by grouping states of similar energies into a single level with a degeneracy given by the density of states at that energy. Along with state grouping, it must be assumed that energy is rapidly randomized among the vibrational levels at a given total energy; this is the rapid intramolecular vibrational relaxation assumption (IVR). Solutions of various Master Equation models have all shown that excitation in the quasicontinuum depends only on laser fluence; this is consistent with numerous experimental studies on a wide variety of molecules [Quack, 1978; Grant et al., 1978; Black et al., 1979; Stephenson et al. 1978; Fuss, 1979; Barker, 1980; Lawrance et al., 1981; Francisco et al., 1982]. The incoherent excitation in the quasicontinuum allows the molecule to be excited up to or beyond the reaction threshold level into the third region, i.e., the dissociative continuum.

88

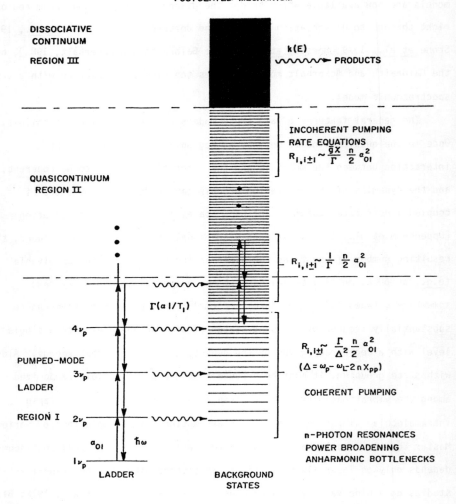

Figure 1.1. Schematic model for infrared multiple-photon excitation, showing transition from coherent pumping in the pump mode ladder (region I) to incoherent pumping in the quasicontinuum (region II) to dissociation in the true continuum (region III). [From J.I. Steinfeld, Molecules and Radiation, 2nd ed., MIT Press, Cambridge, Mass., 1985. Reproduced with permission.]

The reaction threshold level (sometimes referred to as the dissociative threshold) is the boundary point between the QC and dissociative continuum. At or beyond this energy, the levels are characterized by an increasing dissociative width. As a consequence, there is competition between further excitation up the vibrational manifold and reaction. Rates of reactions at energies higher than threshold energy for dissociation are best determined from statistical unimolecular rate theories, such as RRKM theory [Robinson and Holbrook, 1972]. Several IRMPD models have incorporated RRKM rate theory to describe the dissociation process [Mukamel and Jortner, 1976; Stone and Goodman, 1979; Galbraith and Ackerhalt, 1981], but the validation of the use of RRKM was confirmed by molecular beam experiments by Grant and co-workers [Grant et al., 1977] who showed that the dissociation rate for SF_6 is consistent with the RRKM unimolecular rate theory from angular and velocity distributions. These results implied that the vibrational energy is randomized over all the states above the reaction threshold level. King and Stephenson (1977), in their study of the IRMPD of $CF_2C\ell_2$ and CF_2Br_2, used a tunable UV laser to measure the vibrational energy distribution of the resulting CF_2 fragment. Their data showed that the vibrational population distribution can be accounted for by a statistical distribution; this was consistent with conclusions drawn from molecular beam experiments. Subsequent experiments substantiated the conclusions that the internal energy distributions are statistical and the chemistry which occurs proceeds via the lowest reaction pathway which can be characterized by RRKM unimolecular reaction rates [Lussier and Steinfeld, 1977; Richardson and Setser, 1977; Hudgens, 1978; Reiser et al., 1979]. One of the underlying assumptions of RRKM theory is that at energies above dissociation all vibrational modes of the molecule are coupled together and the rate at which energy is randomized or redistributed among the vibrational modes is faster than the reaction rate. Recent work [Dai et al., 1985a] has challenged the assumption of complete randomization of energy at or just above the dissociation limit. The question of whether and to what extent internal energy is randomized below the threshold level, i.e., in the QC, has also received intense study during the past several years; both of these questions will be addressed in the following section.

At this point we have only given a brief overview of the IRMPE and IRMPD processes. In subsequent sections we intend to review the status of new experiments which shed some light on some central questions of RRKM versus non-RRKM behavior in molecules excited to high vibrational energies. The various models which describe the photophysics of the discrete and quasicontinuum regions are expounded in some detail in Section 2, and we comment on recent experimental results which have examined the rôle of (all or some) of the vibrational states in the excitation process occuring in the QC. In Section 3, we review the available experimental techniques for probing the IRMPE process and what they in fact reveal. This includes measurement of bulk properties such as fractional decomposition and average energy absorption; and the use of various spectroscopic techniques such as double resonance spectroscopy, Raman, photoionization, and stimulated-emission pumping to determine the populations of various vibrational levels during the IRMPE and IRMPD process. In Section 4, we suggest several possible applications of IRMPD, in determining chemical-kinetic parameters, for chemical synthesis, and most recently in initiating chemical processes at the gas-surface interface.

2. MODELS OF INFRARED MULTIPLE-PHOTON EXCITATION

A. Bloch Equations and Master Equations

i. Bloch and Master Equation models for excitation.

The physical model described in the preceding section can be formulated mathematically in terms of either classical or quantum-mechanical equations. Attempts to describe IRMPE with classical methods and approximations [Lamb, 1977, 1979; Walker and Preston, 1977; Cotter et al., 1978; Narducci et al., 1979;] require potential energy surfaces for the molecule to be available before the equations of motion can be solved. Unfortunately, accurate potential surfaces for polyatomic molecules are generally not available. Most theoretical treatments have been based on quantum-mechanical equations of motion, utilizing either a "pump mode" or "molecular eigenstate" approach.

The molecular eigenstate method starts from the Schrödinger Equation by treating the interaction of the molecule with the laser field as a process that induces transitions between manifolds of exact molecular eigenstates. The first such treatment, presented by Quack [1978], reduced the time-dependent Schrödinger Equation to a time-independent one using the rotating wave approximation. However, solution of the equations is still a formidable problem, because of the large number of molecular eigenstates involved in the calculations. The problem can be simplified by grouping states of nearly equal energy and quantum numbers into a single level or energy shell and considering only transitions between these grouped levels. Several treatments based on a similar approach have appeared in the literature [Scheck and Jortner, 1979; Mukamel, 1979a,b,c; Carmeli and Nitzan, 1980a,b; Quack, 1982]. In the "pump mode" method, one vibrational mode is considered to interact directly with the laser field, while the remaining vibrational modes act as a heatbath which interacts with the pumped mode through various coupling mechanisms such as anharmonic coupling, dephasing and intramolecular energy redistribution. The theoretical treatments of Mukamel and Jortner [1976a,b], Hodgkinson and Briggs [1976a,b], Cantrell, Galbraith and Ackerhalt [1979], Stone and Goodman [1979], Stephenson et al. [1979], Cantrell [1979], Friedmann and Ahiman [1980], and Galbraith and Ackerhalt [1981] are all based on this approach.

Most of the treatments derive the generalized master equation from a set of optical Bloch equation [Bloch, 1957; Ackerhalt and Shore, 1977] given by

$$i\frac{\partial}{\partial t}\, \underline{\varrho}(t) = \hbar^{-1}[(\mathcal{H}_0 + V),\, \underline{\varrho}(t)] + \underline{\Gamma}\, \underline{\varrho}(t) \tag{2.1}$$

where $\underline{\varrho}$ is the molecular density matrix, $\mathcal{H}_0$ is the molecular Hamiltonian, V is the coherent radiative coupling term, and $\underline{\Gamma}$ is a relaxation operator. In the dipole approximation, V is given by

$$V_{ij}(t) = \mu_{ij}\, E(t) \tag{2.2}$$

where $E(t)$ is the electric field of the laser and μ_{ij} is the dipole transition element. The equations of motion for diagonal elements of the density matrix $\underline{\varrho}$, describing the laser interaction with successive levels along the excitation ladder coupled to bath states, is obtained from Eq. (2.1) as

$$\frac{\partial}{\partial t}\, \rho_{mm} = -\frac{i}{2}\, \Omega_m\, (\rho_{m-1,m} - \rho_{m,m-1}) + \frac{i}{2}\, \Omega_{m+1}\, (\rho_{m,m+1} - \rho_{m+1,m})$$

$$- \left(\sum_{k \neq m} \frac{1}{\tau_{mk}} + \Gamma_m\right)\rho_{mm} + \sum_{k \neq m} \frac{1}{\tau_{mk}}\, \rho_{kk} \tag{2.3}$$

where $1/\tau_{km}$ is the direct non-radiative coupling between levels m and k, Ω_m is the Rabi frequency for the transition m, and Γ_m are rates of dissipation of the level m. Off diagonal elements are given by

$$\frac{\partial}{\partial t} \rho_{mn} = \frac{i}{2} \Omega_n \rho_{m,n-1} - \frac{i}{2} \Omega_m \rho_{m-1,n} + \frac{i}{2} \Omega_{n+1} \rho_{m,n+1}$$

$$- \frac{i}{2} \Omega_{m+1} \rho_{m+1,n} - (i \Delta_{mn}) \rho_{nm}$$

$$- \left(\sum_{k \neq m} \frac{1}{\tau_{mk}} + \sum_{k \neq n} \frac{1}{\tau_{nk}} + \Gamma_m + \Gamma_n \right) \rho_{nm} \tag{2.4}$$

where m<n and Δ_{mn} refers to the laser detuning from the transition between m and n.

To derive rate equations from the Bloch equation (2.3) and (2.4), we must assume that equilibrium is achieved on a time scale much shorter than that needed to pump populations to significant levels of excitation; this implies that $\Omega/W \to 0$, where W refers to the total damping rate. The lowering of the pumping rate relative to the damping rate is produced by a decrease of the Rabi frequency, by an increase of the degree of detuning between the laser frequency and frequency of the pumped mode, or by increase of the damping rate constants. Under these conditions, we may make the Wilcox-Lamb [1960] approximation,

$$\frac{\partial}{\partial t} \rho_{mn} = 0 \qquad (m<n) \tag{2.5}$$

Setting Eq. (2.4) to zero, solving for the off-diagonal elements, and substituting the result into Eq. (2.3), eliminates the off-diagonal elements which appear in the expression for the diagonal equation. This allows the equation of motion for the diagonal components to be written in terms of diagonal elements only, i.e., as rate equations.

$$\frac{\partial}{\partial t} \rho_{mm} = \sum_{k \neq m} R_{mk} \rho_{kk} - \rho_{mm} \sum_{k \neq m} R_{km} + \sum_{k \neq m} \frac{1}{\tau_{mk}} \rho_{kk} - \rho_{mm} \sum_{k \neq m} \frac{1}{\tau_{mk}} - \Gamma_m \rho_{mm} \tag{2.6}$$

where R_{mk} give the radiative transition rates from level m to level k, for which expressions are given by Stone and Goodman [1979], Hodgkinson and Briggs [1979], and Kay [1981]. Equation (2.6) becomes a meaningful rate equation for populations when $P_m = \rho_{mm}$ which yields the phenomenological master equation,

$$\frac{\partial}{\partial t} P_m = L_{m,m-1} P_{m-1} + L_{m,m+1} P_{m+1} - (L_{m-1,m} + L_{m+1,m}) P_m - k_m P_m \quad (2.7)$$

where $L_{m,m-1}$ and $L_{m-1,m}$ are expressions for the rates of absorption and stimulated emission, respectively, if it is assumed that the radiative transition rates R_{km} only couple states differing by one quantum of the laser frequency, i.e., k-m = ±1. The coefficients $L_{mm'}$ have the form of a generalized Lorentzian, including both $R_{mm'}$ and the relaxation rates associated with levels m and m'; a phenomenological reaction rate k_m, out of levels with m > $E_0/h\nu$, has also been included in Eq. (2.7).

ii. Solutions of Bloch and Master Equations

Master equations of the form given in Eq. (2.6) or (2.7) cannot be solved explicitly to yield analytical solutions unless several approximations and assumptions are invoked to facilitate a closed form solution. Yablonovitch and coworkers [Shultz et al., 1978; Black et al., 1979] formulated closed form solutions to the master equations by assuming that the extent of decomposition is small, thereby neglecting the reaction term. Furthermore, in order to approximate first order rate constants for absorption and stimulated emission for level to level transitions given by

$$L_{i+1,i} = \frac{g_{i+1}}{g_i} \quad L_{i,i+1} \quad (2.8)$$

where

$$L_{i,i+1} = \sigma(E_i) \ I/h\nu \quad (2.9)$$

the authors assumed simple functional behaviors for the degeneracies, g_i, and the energy dependent absorption cross-section, $\sigma(E_i)$. The populations resulting from the simple closed form solutions are very similar to a thermal distribution. However, the major limitation of the Yablonovitch treatment is that the method of approximating the degeneracies produces nonphysical results, which raises serious questions concerning the range of validity of the solutions. Fuss [1979] formulated a closed form solution to the master equation by reconstructing Eq. (2.7) into a difference equation in fluence space. This is achieved first by neglecting the reaction term and then changing variables using the fact that $IdT = d\phi$. The difference equation is given by

$$\frac{dP_i}{d\phi} = \frac{1}{g_i} \Delta(g_{i-1}\frac{\sigma_{i-1}}{h\nu}) \Delta P_{i-1} \tag{2.10}$$

where Δ is defined by $\Delta P_i = P_{i+1} - P_i$. If high excitation is assumed, then the discrete level formalism can be replaced by a partial differential equation

$$\frac{\partial P_\varepsilon}{\partial \phi} = \frac{1}{g_\varepsilon} \frac{\partial}{\partial \varepsilon} (g_\varepsilon \frac{\sigma_\varepsilon}{h\nu} \frac{\partial P_\varepsilon}{\partial \varepsilon}) \tag{2.11}$$

However, analytical closed form solutions can only be obtained from the above equation for special cases where the cross-sections and degeneracies assume power function dependences, namely,

$$\sigma_\varepsilon = \sigma_0(\varepsilon + c)^\alpha$$
$$g_\varepsilon = g_0(\varepsilon + c)^{s-1}$$

where σ_0, g_0, c, α, and s are constants; α is constrained to values less than 2. The resulting closed form solutions, however, do not describe the time evolving population distribution, but only provide the net dissociation probability at a specified fluence. The dissociation probability predicted by the Fuss model overestimates the experimental dissociation data in most cases. Medvedev [1979] has provided analytical expressions for the time evolving populations for limiting case where the pump rate is greater than the emission rate, $L_{i,i+1} \gg L_{i,i-1}$, and under the condition that the populations in the lower levels are quasistationary. These assumptions give rise to the general expression for the population in level i, at time t, as

$$P_i\ (t) = \frac{1}{(i-1)!}\ \theta^{i-1}\ \exp\ (\theta) \tag{2.12}$$

where $\theta = (I\sigma t/h\nu)$. The form of the population distribution is almost that of a Poisson distribution. Troe [1980] derived general expressions for three limiting cases: case A, the weak field limit, which considers the radiative transition rate and emission rates above the critical threshold to be slower than the rate of reaction; case B, in which the rate of reaction is smaller than the pumping and emission rates above the threshold; and case C assumes the initial distribution to be at equilibrium. Results from the general population expression given by Troe usually do not give good agreement with populations calculated from exact numerical solutions. Dilonardo [1980] has modified Troe's treatment for weak and moderate intensity fields, but the calculated results only give qualitative information and at best only provide semiquantitative analysis of IRMPD experimental data.

A general solution to the master equation can be obtained numerically. Numerical solutions are far more exact and extendable over a larger range of conditions of practical interest than the derived analytic solutions, albeit the computer time requirements are large. Several methods are now available which use various numerical procedures such as Gear integration, Monte Carlo, or matrix methods. Numerical integration procedures that employ the Gear integration method have been used by several researchers [Lyman, 1977; Eleckij et al., 1977; Grant et al., 1978; Schulz et al., 1980; Dai et al., 1980].

Direct integration of the master equation need not incorporate limiting assumptions to simplify the form of Eq. (2.7). In modeling the IRMPE process for most systems, a large number of levels may need to be considered (ca. 30 or more levels for chloroethane) which implies that a large number of coupled stiff differential equations must be integrated; this could require large amounts of computer time. Barker [1980] formulated an alternative numerical approach which reduces the amount of computer time required. His procedure is based on the Monte Carlo stochastic method using the Gillespie algorithm [Gillespie, 1976, 1977]. The stochastic treatment assumes that the chance that a molecule will undergo stimulated emission, stimulated absorption, or unimolecular reaction during the laser pulse can be described by a joint probability distribution function assuming that the molecule is in a particular state, i.e., ground state. Given that the molecule is well defined a random time interval can be generated and the history of the molecule can be followed through a series of randomly selected events (absorption, emission, and reaction). The procedure is repeated until decomposition takes place and suitable statistics are achieved for the ensemble of molecules. Although this method avoids integration of stiff differential equation which consumes computer time, it too requires ample computer resources when detailed information such as populations are desired because of the large number of computed trajectories needed to obtain good accuracy, but few trajectories are needed to calculate experimental bulk average quantities such as fractional decomposition and average number of photons absorbed. Weston [1982] used the stochastic method in modeling IRMPD data for CF_3I, CF_3Cl, and CF_3CH_3. He found the procedure adequate in predicting bulk average quantities, but more subtle features of the IRMPE process, such as the population distributions, could not be described properly.

Numerical solution of the master equation based on a matrix formulation developed by Lawrance et al. [1981] entails diagonalizing a symmetric tridiagonal matrix which embeds terms for absorption, emission and reaction during the pulse. The resulting eigenvector matrix is inverted and then used in several matrix multiplication operations to obtain the time dependent population vector. Lawrance and co-workers also modified the matrix procedure to include reaction following the laser pulse and post-pulse collisions. Reaction following the pulse is particularly important for molecules which possess low reaction threshold energies and slow decomposition rates, or for molecules excited by short laser pulses. In these cases post-pulse dissociation can be the dominant contribution to measured dissociation yields. Most other numerical or analytic solutions of the master equation, to date, seem to either truncate or neglect subsequent reaction following the laser pulse and most often ignore post-pulse collisions. Eberhardt et al. [1982] successfully used this procedure to model results for ethyl acetate which illustrate the importance of post pulse effects.

iii. Comparison of Theory with Experiment

In comparing theoretical models of IRMPE with experimental data, it is essential that the measurements and calculations refer to the same quantity. Bloch and Master Equation calculations are generally performed for uniform laser intensity or fluence, while experimental results frequently are derived from focussed laser beams having nonuniform intensity throughout the beam. Thus, the intrinsic distribution of responses at a given intensity is further convoluted with a distribution of intensities. Consequently, before any interpretation of IRMPE and IRMPD data can be made, and compared with physical models, the data must be corrected for the beam geometry.

The problem is essentially one of either convoluting theoretical predictions with an intensity distribution to obtain a predicted experimental observable, or deconvoluting the experimental data to obtain microscopic values. For Gaussian beams, several methods of carrying out this procedure have appeared in the literature [Gower and Billman, 1977a; Fuss and Cotter, 1977; Lyman et al., 1977; Kolodner et al., 1979; Black et al., 1979; Reiser and Steinfeld, 1980; Francisco et al., 1981]. In most of these treatments a Gaussian beam is assumed which has an intensity profile along the beam

$$I(r,z) = I_0 \ [w_0^2/w^2(z)] \ \exp \ [-2r^2/w^2(z)] \tag{2.13}$$

where r is the radial distance, z is the coordinate along the beam axis,

$$w(z) = w_0 \ (1+z^2/z_0^2)^{1/2} \tag{2.14}$$

and $z_0 = \pi w_0^2/\lambda$, with λ being the wavelength and w_0 the beam waist. However, in most IRMPE experiments and theories, fluence

$$\phi(r,z) = \int_{pulse} I(z,t)dt$$

is the primary variable. The spatial dependence of the fluence for a Gaussian beam has the same functional form as the intensity profile, namely,

$$\phi(r,z) = \phi_0[w_0^2/w^2(z)] \ \exp \ [-2r^2/w^2(z)] \tag{2.15}$$

In any experiment, the net physical parameter or observable $G(\Phi)$ that is measured represents the sum of the responses of the system over all portions of the beam; Φ is the average fluence over the beam. This is expressed by the integral equation

$$G(\Phi) = 2\pi \int_0^L dz \int_0^\infty dr \ r \ g(\phi(r,z)) \tag{2.16}$$

where the integration volume is over the length of the cell L, and since the fluence goes to zero well inside the inner diameter of the cell, the radial integral can be conveniently taken to $r \rightarrow \infty$.

The object of deconvolution is to extract the geometry independent observable, $g(\phi(r,z))$ for a constant (fixed) fluence. The treatment presented by Kolodner et al. [1979], formulates a deconvolution procedure for the case of a nearly collimated gaussian beam of radius w_0 beginning from Eq. (2.16).

Under these conditions, $z \approx z_0$ implies a uniform beam throughout the cell. It is further assumed that the optical depth of the medium is small. With these approximations, the integral equation given by Eq. (2.16) can be rewritten as

$$G(\Phi) = \int_0^{\phi_0} g(\phi) \frac{d\phi}{\phi} \tag{2.17}$$

The physical response function, $g(\phi)$, is obtained by differentiating both sides of Eq. (2.17) with respect to ϕ_0 and changing variables, which yields

$$g(\phi) = \phi \frac{dG(\phi)}{d\phi} = G(\phi) \frac{d\ln G(\phi)}{d\ln \phi} \tag{2.18}$$

This approximation procedure as originally pointed out by Kolodner et al. [1979] only holds for physical observables such as average energy and total dissociation probabilities. But more importantly it is valid only for treatment of data taken under collimated beam conditions rather than focused beam conditions.

A more general derivation of the deconvolution equation for focused Gaussian beams was given by Reiser and Steinfeld [1980]. In their treatment it is assumed that molecules which undergo irradiation are contained in the focal volume of the laser beam. Keefer et al. [1976] obtained an expression for the volume enclosed by an iso-intensity, hence, iso-fluence surface as

$$V(\phi) = -\frac{\pi^2 w_0^4}{\lambda} \int_0^{u'} (1+u^2) \ln[\kappa(1+u^2)] du \tag{2.19}$$

where $\kappa = \phi/\phi_0$

          $u' = (1/\kappa - 1)^{3/2}$

and $u = (z-z(0))/z_0$

The reaction volume $U_R = V(\phi > \phi^*)$, where ϕ^* is the minimum fluence required for dissociation, is found to be proportional to $\Phi^{3/2}$. Early IRMPD studies of SF_6 showed this 3/2-power dependence on fluence, which was initially attributed to microscopic effects such as dynamic Stark broadening. Subsequently, however, the geometric origin of this dependence was recognized [Keefer et al., 1976; Speiser and Jortner, 1976].

For a focused beam the general deconvolution equation is given by

$$g(\phi) = \frac{E_T}{n\phi v(\phi)} \left(\frac{\partial G(\phi)}{\partial E_T} \right) \tag{2.20}$$

where n is the number density of absorbing molecules, E_T represents the total pulse energy, and $v(\phi)$ is the differential volume, i.e., change in volume per unit change in ϕ given as

$$v(\phi) = \frac{dV(\phi)}{d\phi} = \frac{-\lambda z_0^2}{3} \left(\frac{u(2\kappa+1)}{\phi_0 \kappa^2} \right) \tag{2.21}$$

with the condition that $\kappa < 1$ for focused beams.

In the various deconvolution methods, derivatives must be evaluated, and are usually obtained from the slopes of the total response curve with fluence or total energy. This entails determining local derivatives which are often extremely noisy since they are obtained from small differences taken from large numbers. Unfortunately, the approach yields unreliable results, but this problem can be circumvented by fitting the experimentally determined curve with an algebraic function which allows derivatives to be obtained analytically [Black et al., 1979]. Francisco et al. [1981] presented an alternative method which allows deconvolutions to be done using a least-squares method. Despite these various methods of improving the certainty in deconvoluting experimental data, there are several problems inherent in deconvoluting data [Reiser and Steinfeld, 1980; Francisco et al., 1981]. First, small IRMPE dynamical effects are completely smoothed out in the deconvolution process. Second, as is more often the case, the observed data are usually sparse; this, coupled with the low experimental precision, ultimately limits the value and reliability of deconvoluted data.

Alternatively, if a meaningful comparison between theoretical and experimental results is to be made, convolution of the theoretical derived response function, $g(\phi)$ is a better procedure, because data can be easily obtained to define the functional dependence of the observable. The spatially averaged observable (response function) is derived from the integral equation (2.16). The double integral can be solved numerically over the irradiation volume. Gower and Billman [1977a] have used this approach, but it does require considerable computer time. Francisco et al. [1981] devised a procedure which allows evaluation of Eq. (2.16) analytically. This is done by assuming that $g(\phi)$ can be represented functionally as a power series in ϕ,

$$g(\phi) = \sum_{n=1}^{\infty} a_n \phi^n \qquad (2.22)$$

subject to the constraint that the function approaches zero as $\phi \to 0$; the a_n are determined from the data by least-square fitting. With Eq. (2.22) substituted into (2.17), and normalizing over the entire volume of the cell V_T, $G(\Phi)$ can be obtained from the analytical expression

$$G(\Phi) = V_T^{-1} \sum_{n=1}^{\infty} n^{-1} \left(\frac{w_0}{\lambda}\right)^{2n-2} a_n \pi^{n-1} z^n I_n E^n \qquad (2.23)$$

where the total pulse energy is

$$E = \frac{1}{2}\phi_0 \pi w_0^2 \qquad (2.24)$$

The I_n integrals may be found in Francisco et al. [1981].

B. Quasicontinuum Model, Intramolecular Vibrational Relaxation, and
       Quantum Chaos

Before proceeding to the discussion of experimental probes for IRMPE dynamics, let us pause for a moment to re-examine the key assumption required to obtain a description in terms of incoherent rate equations, namely, rapid redistribution of the excitation energy among all possible vibrational modes of the molecule. As shown in Table 2.1 [Thiele et al., 1980], if this intramolecular vibrational relaxation (IVR) process is fast compared with either the IR pumping rate or the rate of decay by reaction, then the molecule will tend to react statistically, i.e., all reaction channels will be open and the likelihood of each channel will be proportional to its associated phase space. An important consequence of this is that the overall chemical behavior will be well described by RRKM unimolecular kinetics [Robinson and Holbrook, 1972].

The experimental evidence overwhelmingly supports the rapid IVR assumption. Essentially all studies of product distributions in gas-phase IRMPD can be satisfactorily explained using the Master Equation/RRKM model. The single notable exception is the bimolecular rate enhancement in low-temperature matrices studied by Pimentel and co-workers, which is discussed in Section 4.A(iii). Since in this work the observed mode selectivity occurs at a total energy content of ca. 2000 cm^{-1}, IVR in the quasicontinuum is not an issue.

Spectroscopic measurements on vibrationally excited molecules also support the rapid IVR assumption. IR fluorescence data from dimethyl ether and 1,4-dioxane have been interpreted in terms of statistical energy redistribution [Kulp et al., 1984, 1985; Stewart et al., 1984] as have optical fluorescence measurements on the S_1 states of stilbene and anthracene [Felker and Zewail, 1985 a,b,c; Felker et al., 1985]. One type of experiment that has received wide publicity as conclusive evidence for rapid IVR is the measurement of the widths of overtone absorption bands, e.g. of benzene and its deuterated analogues [Reddy et al., 1982; Sibert III et al., 1984 a,b].

In these studies, the width of the band is assumed to be due only to homogeneous broadening, and the "lifetime" is obtained simply from

$$\tau = (2\pi c \Gamma)^{-1} \tag{2.25}$$

where Γ is the observed width in cm^{-1}. This results in lifetimes on the order of 10^{-13} sec.

Several recent experiments have shown that this interpretation is grossly oversimplified and may be seriously in error. Heilweil et al. [1985] have measured directly the T_1 relaxation times of surface OH chemisorbed on SiO_2 using picosecond pump/probe techniques, and have found that the "lifetime" derived from the width of the associated absorption band may overestimate the actual lifetime by one or two orders of magnitude. Dübal and Crim [1985] have studied the overtone spectra of H_2O_2, and have found that these bands are mostly inhomogeneously broadened, with a rich rotational fine structure.

Of course, linewidths can be used to estimate lifetimes using Eq. (2.25) if the individual rotational features are resolved, as in the experiments of Dai et al. [1985a] on formaldehyde. In this study, linewidths of 10^{-4} cm^{-1} are observed, leading to lifetimes in the 10^{-8} sec range; what is notable is that of two rovibrational levels near the top of the potential barrier, the one with 31 cm^{-1} higher total energy has a 2.5 times slower decay rate, suggesting that a simple RRKM model may not be applicable.

There are several factors which may block rapid IVR, and thus make mode-selective processes possible. First, there may be symmetry restrictions on the effective density of vibrational states within the molecule, which would limit the range of states into which the energy can be dissipated [Stone and Goodman, 1985; Sinha and Kinsey, 1983]. There also appears to be to be strong rotational-state effects on intermode coupling [Kyrölä and Eberly, 1985]. Evidence for such effects have been noted in IR fluorescence [Kulp et al., 1984, 1985; Stewart et al., 1984], optical fluorescence [Felker and Zewail, 1985 a,b,c; Felker et al., 1985] and stimulated-emission pumping experiments [Dai et al., 1985b]; as well as in several of the double-resonance experiments described below in Section 3.B.

Table 2.1. Rôle of IVR in Determining Mode Selectivity
[from Thiele et al., 1980. Reproduced with permission].

random distribution of states for given energy

reaction is described by a single RRKM decay constant for given energy

or

NO SELECTIVITY

fast **IVR/LASER PUMP RATE** slow

non-random distribution of states for given energy

fast **IVR/DECAY RATE** slow

reaction not described by a single decay constant for given energy, intrinsically non-RRKM

and

SELECTIVITY POSSIBLE

excitation of states with energy high above E_a is impossible

fast **DECAY RATE/LASER PUMP RATE** slow

excitation of states with energy high above E_a is possible

THERMAL REACTION PRODUCTS

PSEUDO-SELECTIVITY NON-THERMAL REACTION PRODUCTS POSSIBLE

3. TECHNIQUES AVAILABLE FOR PROBING THE IRMPE PROCESS

In numerous experimental studies of IRMPA and IRMPD, several methods have been used to interrogate the three regions of the multiple-photon excitation process delineated in Figure 1.1. The most desirable experimental result would be the complete time evolution of the populations, and how they depend on laser power, laser fluence, and the properties of the molecule. Until recently, such measurements have been difficult to obtain. Direct measurements of population distributions, derived from induced absorption (double-resonance spectroscopy), spontaneous and induced emission, or photoionization, can yield information on absorption of the first few photons, the rôle of compensation mechanisms such as anharmonic coupling or rotational hole filling in allowing additional photons to be absorbed, and the optical laser parameters which influence the evolution of the populations. On the other hand, indirect measurements such as fractional decomposition, average absorption cross sections, and average energy deposition are highly averaged over the population distribution. Yet, these measurements are just as essential as direct spectroscopic observations to understanding the ·multiple photon process. Direct probes of Region 3, which lies above the dissociation threshold, are unavailable for most molecules; yet product yields from dissociation or isomerization reactions can probe the high energy tail of the vibrational energy distribution. The qualitative and quantitative information deduced from such measurements, and how this information defines critical properties of molecules which undergo IRMPE, is reviewed in the first part of this section. We also consider the extent to which detailed dynamical information can be extracted from such data. We then consider direct probes of population distributions by spectroscopic techniques, from which the dynamics of the IRMPE process can be elucidated.

A. Bulk Measurements

The first studies of IRMPD relied on the chemical behavior of molecules as a method of interrogating time evolving populations. The chemical systems used were those with simple and clean photochemistry, which could easily be deduced from the net chemical products. The essential requirements of simple and well-defined chemistry are important for obtaining accurate and meaningful measurements of key experimental observables such as product yields, branching ratios, energy deposition, and microscopic absorption cross sections.

i. Product yield and branching ratio measurements

To extract information on the dynamics of multiple-photon excitation in competition with reaction above the dissociation threshold, and some details regarding the excitation of the molecules prior to dissociation, from product yields and branching ratios, it is essential that such observables are accurate and well-understood functions of laser frequency, intensity, and fluence.

Both dissociation yields and branching ratios require quantitative measurements of products from IRMPD of reactant molecules. These measurements may be carried out in molecular beams or in static gas cells. In a molecular beam, true collisionless conditions may be attained, while most experiments in static gas cells are carried out under the influence of molecular collisions. If pressures are kept sufficiently low (< 10 Pascals) so that the mean collision time is greater than the duration of the CO_2 laser pulse, collisional effects during excitation may be minimized; however, post-pulse collisions may modify population distributions and thus product yields.

Typical IRMPD experiments involve irradiating samples of reactant gas with a pulsed infrared laser and detecting its products. Standard analytical techniques are used to detect dissociation products in molecular beams and static gas cells. Techniques used in molecular beams include electron impact ionization and chemi-ionization [Schulz et al., 1979], monitoring photolytic depletion of the molecular beam [Brunner et al., 1977], and laser-induced fluorescence [Campbell et al., 1976]. Those that have been used in static gas cells are summarized in Table 3.1. The methods which rely on optical or mass spectrometry provide real-time determination of primary photolysis products and allow the reaction dynamics of the dissociating system to be monitored directly from individual laser pulses, but may be difficult to calibrate. Other techniques capable of furnishing absolute product yields are infrared spectroscopy, gas chromatography, and pressure changes upon irradiation; but these methods integrate over all pulses. Nevertheless, these techniques are generally simpler and more readily available and require no special knowledge of optical properties of reactant or product molecules. These techniques are also employed in the measurement of branching ratios, since these are ratios of product yields from individual channels in molecules with multiple reaction pathways. Products from each channel are usually chemically distinct and can be measured and characterized unambiguously.

If we are to understand the IRMPA and IRMPD processes, we need to know how the various molecular and laser properties influence experimental results such as fractional yields and branching ratios. Fractional decomposition yields are given by the fractional population above the critical energy E_0, which can be obtained from the time evolving populations as

$$f = 1 - \left\{ \sum_{i=1}^{\infty} P_i(t) / \sum_{i=1}^{\infty} P_i(0) \right\} \tag{3.1}$$

where $P_i(0)$ and $P_i(t)$ are initial and final populations, respectively, for energy state i. The effects of laser fluence, laser intensity, and vibrational density of states are implicit in the population distribution.

Table 3.1

Some Techniques Used to Measure Yields in Static Systems in IRMPD Studies

Techniques	Type of Chemical System Studied	Species Detected	Reference
luminescence from new species	SiF_4	SiF_2, SiF	Isenor et al., 1973
	C_2H_3F, C_2H_5F, $C_2H_2F_2$, $C_2H_4F_2$	HF	Quick and Wittig, 1978
	$C_2H_3F_2C\ell$	HF	West et al., 1978
	$C_6H_{12}O_2$	C_3H_6O	Yahav and Haas, 1978
laser-induced fluorescence from products	CH_3NH_2	NH_2	Hancock et al., 1979
	$CF_2HC\ell$	CF_2	Stephenson et al., 1979
	C_2H_4	C_2	Chekalin et al., 1977
	$(CH_3CO)_2O$	CH_2	Ashfold et al., 1979
	$CH_2=C=O$	CH_2	Grimley and Stephenson, 1981
	CH_3ONO	NO	King and Stephenson, 1985
UV absorption spectroscopy	$CF_2HC\ell$	CF_2	Duperrex and van den Bergh, 1979
	O_3	O_3^*	Proch and Schroder, 1979
infrared spectroscopy	SF_6	SF_6	Ambartzumian et al., 1976
	SF_6	SF_6	Kolodner et al., 1977
	$CF_2HC\ell$	$CF_2HC\ell$,C_2F_4	Duperrex and van den Bergh, 1979
	CF_3OOCF_3	CF_2O,CF_3OF	Francisco et al., 1981b
gas chromatography	HCOOH	CO,H_2O,CO_2,H_2	Corkum et al., 1977
	C_4H_8O	C_2H_2,C_2H_4O, C_2H_4,C_4H_{10}	Brenner, 1978
	$C_2H_5C\ell$	C_2H_4	Dai et al., 1980
	$C_4H_5O_2$, $C_5H_7O_2$	$C_4H_4O_2$,$C_5H_7O_2$, C_2H_4	Danen et al., 1982
mass spectroscopy	$CF_3CDC\ell_2$, CDF_3, $CHDF_2$	C_2F_2H,C_2F_2D	Marling et al., 1980
	CF_3COCF_3	C_2F_6,CO	Hackett et al., 1979

These parameters control the initial flow of population from lower levels to higher ones, and consequently contribute to population movement in the QC and above threshold. To understand how these parameters influence fractional yield data, numerous experimental studies have characterized the functional dependences of fractional yields versus laser pulse shape (intensity), laser fluence, and laser frequencies for various molecules. Yields are sometimes expressed as reaction or dissociation probabilities, the values of which may range from 10^{-6} to 1. The yield magnitude and how rapidly the yields saturate with laser fluence provide qualitative information on the influence of early bottlenecks associated with anharmonicity effects in the dissociation region. In ethyl acetate reaction yields at high fluences approach 100%, meaning all molecules within the irradiation region undergo complete reaction [Danen et al., 1982]. In molecules such as hexafluoroacetone, limiting yields were less than 50% [Fuss et al., 1979]. This implies that molecules with fewer initial discrete levels are less effectively pumped to and above the reaction threshold.

There have also been a number of studies which have examined the frequency dependence of reaction yield data. Red shifts of the absorption spectrum with decreasing photon energy have been observed, due to anharmonicity in lower discrete levels. Fuss et al. [1979] demonstrated this in IRMPD experiments on SF_6. Similar conclusions were also reached from studies on chloroethane by Francisco et al. [1983], in which red shifts of 5-6 cm^{-1} were observed. Results on this molecule using a 3.3 μm excitation source revealed no apparent shifts [Dai et al., 1979]. This suggests that anharmonic compensation is most significant at lower levels; when the quasicontinuum is reached (by use of higher-energy photons or by successive perturbation of the molecule), resonant absorption occurs more easily. Red shifts of 30 cm^{-1} have been observed in hexafluoracetone [Fuss et al., 1979].

Branching ratio data, on the other hand, can provide information on the competition between further excitation and intramolecular or intermolecular relaxation above the lowest critical energy threshold for reaction. Several chemical systems have been used to explore the rôle of these competing effects. Some examples are ethyl vinyl ether [Brenner, 1978; Rosenfeld et al., 1977], $CF_2C\ell Br$ [Morrison et al., 1981], $C_2H_4FC\ell$ [Baklanov et al., 1979; Richardson and Setser, 1977], $C_2H_2FC\ell$ [Jalenak and Nogar, 1980], cyclobutyl chloride [Francisco and Steinfeld, 1981], vinyl chloride [Reiser et al., 1979], and vinylcyclopropane [Farneth et al., 1979]. In the dissociation of ethyl vinyl ether, Brenner [1978] found that shorter laser pulses gave more products resulting from higher energy reaction channels than longer pulses. This result showed that the faster excitation rates of shorter pulses compete with dissociation by the lowest energy reaction channel. Rosenfeld et al. [1977] studied the IRMPD of the same system at pressures between 5 and 44 Torr and were able to explore the role of collisions in redistributing energy above the lowest reaction threshold. They found that branching ratios were insensitive to changes in the reactant pressure; and consequently, it was concluded that a thermalized system resulted at these high pressures and, assuming the applicability of RRKM theory, a population temperature was estimated for the system to be ~1600 K. Results from these two experiments suggest that if the initial population of excited molecules produced by IRMPA is initially non-Bolzmann above the threshold, then collisional processes thermalize the system rapidly and efficiently.

ii. Average absorption cross section and average energy deposition measurements

In addition to product yields and branching ratios, effective

macroscopic absorption cross section and average energy deposited in a
molecule may be used to characterize absorption properties of excited
molecules. These measurements may be determined by two methods: laser
attenuation and optoacoustics. The effective laser absorption cross section,
$\sigma_L(\Phi)$, as measured from laser attenuation experiments is given by

$$\Phi/\Phi_0 = \exp[-\sigma_L(\Phi)N\ell] \tag{3.2}$$

when Φ_0 is the incident laser fluence; Φ, the transmitted fluence through the
sample cell of length ℓ and containing N molecules per cm^3 in the irradiation
volume. The value of $\sigma_L(\Phi)$ is sensitive to several factors, including
excitation frequency, optical beam profile, focusing geometry, pulse length,
temperature and pressure-broadening effects. Furthermore, these cross
sections must be measured over a range of laser fluences which do not exceed
the reaction threshold fluence. Laser attenuation methods, albeit more
direct, are limited to molecules which strongly absorb at the excitation
wavelength. For weakly absorbing systems, the gas cell must be optically
thick, but this leads to varying fluences along the cell which obscures direct
data interpretation. Nevertheless, from the macroscopic absorption cross
sections, $\sigma_L(\Phi)$, and the incident laser fluence, Φ_0, the total energy absorbed
can be deduced from,

$$\langle E|\Phi\rangle = \Phi_0\sigma_L(\Phi) \tag{3.3}$$

and dividing by the photon energy gives the average number of photons absorbed
per molecule,

$$\langle n\rangle = \frac{\Phi_0\sigma_L(\Phi)}{h\nu} \tag{3.4}$$

Another technique used to measure average energy absorbed in a molecule
undergoing IRMPE is the optoacoustic method. This method is sensitive and has
greater application to weakly absorbing molecules and low pressures than the
laser attenuation method. Burak et al. [1970] first observed acoustic waves
in laser-pumped SF_6; Bagratashvili et al. [1976] first proposed using the

optoacoustic technique as a direct way of monitoring energy deposition in SF_6 and OsO_4. This technique has been applied to other polyatomic molecules as well [Deutsch, 1977; Lussier et al., 1978; Black et al., 1979; Kwok et al., 1981; Stephenson et al., 1982; Danen et al., 1982; Presser et al., 1983; Bloembergen et al., 1984; Ashfold et al., 1984; Evans et al., 1984]. The technique is based upon the detection of acoustic waves generated by expansion of the gaseous sample upon absorption of energy from the laser. The gas expansion results from a temperature rise in the interaction region due to relaxation of the vibrational excitation in the molecule into translational energy which is accompanied by a pressure increase. The pressure gradient is detected by a very sensitive microphone. Assuming complete thermalization of the absorbed energy, the amplitude of the acoustic signal generated by the pressure gradient can be assumed to be directly proportional to the energy absorbed. Absolute calibration of the optoacoustic signal is made by comparison with laser attenuation measurements using either a well defined collimated Gaussian beam, which yields nearly constant fluences over a short path length, or an unfocused beam. The average energy absorbed is related to average number of photons and macroscopic cross section through equations 3.3 and 3.4.

Other techniques which can in principle probe energy absorption in molecules following IRMPE are bolometric measurements [Gough et al., 1978] and thermal lensing measurements [Fang and Swofford, 1983].

To date, there have been numerous measurements of macroscopic absorption cross sections and mean number of photons absorbed as functions of laser fluence, laser intensity, and laser frequency [Ambartzumian et al., 1976c; Ham and Rothschild, 1977; Kwok and Yablonovitch, 1978; Danen et al., 1982; Evans et al, 1982]. Observed trends in the data provide at most only qualitative insight into how absorption in the lower region affects average cross sections and <n>. The various effects of anharmonicity are inferred from a highly averaged and convoluted quantity, as shown in Eq. 3.5, which relates the macroscopic absorption cross-section to the microscopic absorption cross section:

$$\sigma(\Phi) = \frac{\int_0^\infty \sigma(E)P(E,\Phi)dE}{\int_0^\infty P(E,\Phi)dE} \qquad (3.5)$$

where the function $P(E,\Phi)$ is the normalized population distribution prepared at a particular fluence.

iii. Summary of properties derived from bulk measurements

What can be learned from qualitative trends observed in bulk measurements about the IRMPE process? Comparing trends in bulk measured observables and molecular parameters for numerous molecules undergoing IRMPE show that molecules can be grouped according to size: small, intermediate, and large. In each of these categories, yield and absorption measurement dependences on various laser parameters do exhibit consistent trends. Some of these characteristics are summarized in this section for a few molecules. Lyman et al. [1981] have collected and compiled an extensive list of IRMPA, IRMPD, and molecular properties for a larger set of molecules.

The dependence of measured product yields on parameters such as laser fluence or intensity provides information on the importance of coherent vs. incoherent laser excitation. For small molecules with relatively few atoms (i.e., 3 to 4 atoms) characterized by discrete vibrational levels, the effect of excitation through the discrete levels on IRMPA and IRMPD has been found to depend on laser intensity. According to theory [Stone and Goodman, 1979], the excitation should be dominated by intensity effects. Experimental studies on triatomic molecules [Burak et al., 1979a; Bloembergen et al.,

1984; Wolk et al., 1980; Proch and Schröder, 1979; Avouris et al., 1979; Hartford, 1978] have demonstrated that induced excitation is determined by laser intensity, as is evident from small dissociation yields. In studies of OCS with CO_2 pulses of 0.5 nsec at an intensity of 200 J/cm^2, dissociation yields per pulse were $\sim 6 \times 10^{-4}$. The average number of photons absorbed exhibited strong intensity effects for CO_2 laser pulses of 0.5 nsec, and $<n>$ was less than unity up to a fluence of 200 J/cm^2 and intensities of 300 GW/cm^2. The experimental trends for OCS, have been observed in other small molecules and strongly suggest that excitation in such molecules is dominated by region I.

In molecules with more than 10 atoms ("large" molecules), where at room temperature the density of vibrational and rotational states may exceed $\sim 10^3/cm^{-1}$, the excitation is governed by incoherent laser pumping through the quasicontinuum, which is dependent on laser fluence. Dissociation yields for these molecules may approach unity under low fluence conditions (1-10 J/cm^2). Experimentally, this has been demonstrated from IRMPD studies on such systems as ethyl acetate [Danen et al., 1981, 1982] and bis(trifluoromethyl)peroxide [Francisco et al., 1981]. However, not all molecules with a large number of atoms exhibit "large molecule" behavior. IRMPD studies on cyclobutyl choride [Francisco and Steinfeld, 1981] and on cis-3,4-dichlorocyclobutene [Ringwelski et al., 1984; Presser et al., 1983a,b] both exhibit intensity and fluence effects on yields and absorption measurements. For these systems, excitation must occur through a few discrete levels which provide a small bottleneck. Molecules with total atom numbers between 4 and 10, may exhibit behavior intermediate between properties of small and large molecules. For these molecules both intensity and fluence effects are important. These effects may arise in several ways: (1) coherent excitation through several low-lying discrete states; (2) competition between

laser-pumping and molecular relaxation caused by collisions; and (3)
competition between chemical reaction and further laser excitation above the
dissociation threshold. Several researchers have found that intermediate
molecules exhibit intensity effects on dissociation yields [Stephenson, 1982;
Stephenson and King, 1983; Dai et al., 1980; Francisco et al., 1981, 1983].
But in the case of SF_6, extensive studies have shown that dissociation yields
are dependent on fluence [Kolodner et al., 1977; Lyman et al., 1977; Gower and
Billman, 1977b], while the average number of photons absorbed per molecule
(energy deposition) is intensity dependent [Black et al., 1979; Kwok et al.,
1981]. For intermediate molecules, red shifts of the absorption spectrum with
frequency are expected because of the anharmonicity due to the vibrational
discrete levels. This can be seen from the measured IRMPD yield spectrum with
CO_2 laser excitation for chloroethane [Francisco et al., 1981, 1983]. A
summary of the behavior of variously sized molecules has been given by
Bloembergen et al. [1984; also Simpson et al., 1985], and is shown in Table
3.2. While this provides a good guide to IRMPE behavior, as with all
generalizations numerous exceptions can be found.

Table 3.2. IRMPE behavior for variously sized molecules [Bloembergen et al., 1984; Simpson et al., 1985. Reproduced with permission]

Diatomics	3 and 4 atom molecules	4-6 atom molecules	6-10 atom molecules	10+ atom molecules
Strict multi-photon excitation	Intensity controls excitation	Both intensity, fluence important	Fluence dominates intensity	Fluence controls excitation
No known examples of IR dissociation	At 100 GW/cm²: -small dissociation yield -small $\langle n \rangle$	Unity dissociation yield, 10-50 J/cm², lower for <1 ns t_p	Unity dissociation yield, 10 J/cm²	Unity dissociation yield, 1-10 J/cm²
Theoretical calc. estimates 10-100 TW/cm² for dissociation	Region one dominates	Large bottleneck — Both region one, quasicontinuum effects	Small bottleneck — Quasicontinuum dominates	No bottleneck — Molecule starts in quasicontinuum

Region One
Excitation via resonantly enhanced high order multiphoton processes to quasicontinuum

Little population of intermediate excited states due to lack of resonances

Quasicontinuum
One-photon, stepwise, incoherent excitation

Broad distribution of final excited states

iv. Energy distributions from bulk measurements

The nature of the internal energy distribution in the molecule after a laser pulse, as mentioned in the introduction, is central to understanding the details of the multiphoton process. Early experimental estimates of the internal energy distribution in the quasicontinuum and the dissociative continuum from bulk measurements were extracted from averaged observables such as dissociation yields, branching ratios, and average energy deposition. From yield data, the population in the tail of the distribution function can be estimated, since molecules with energy above the threshold E_0 will react. The distribution of internal energy below E_0 is deduced from yield data with use of a "frozen" distribution model. This model, originally proposed by Shultz and Yablonovitch (1978), showed that the fractional decomposition is given by

$$f = \int_{E_0}^{\infty} P(E)dE \tag{3.6}$$

which is just the total population of the molecule above the threshold for reaction, assuming that $P(E)$ is statistical. Similarly, branching ratios can be defined as the ratio of average rates for a molecule with multiple reactive channels given by

$$\frac{f_i}{f_j} = \frac{\int_{E_{0,i}}^{\infty} k_i \, P(E)dE}{\int_{E_{0,j}}^{\infty} k_j \, P(E)dE} \tag{3.7}$$

It can be seen from Eq. (3.6) and (3.7) that both fractional dissociation yields and branching ratios sample the same portion of the distribution function. When the differences in critical energies for the lower reaction channels are small, measurements of both branching ratios and yields provide redundant information regarding the form of the distribution. Large critical energy differences are therefore more informative. The forms of distribution that can be extracted from these data have been studied by numerous researchers. Reiser et al. [1979] attempted to extract the distribution function shape from yield and branching ratio data by comparison of those calculated from suitable choices of distributions to the yields and branching ratios measured in IRMPD of vinyl chloride via isomerization and dissociation. The authors found that two postulated distributions, i.e. Poisson and Boltzmann distribution, were capable of reproducing their data, and concluded that on the basis of yields alone, a unique distribution could not be determined. Similar studies on cyclobutyl chloride confirmed these conclusions [Francisco et al., 1981; Lawrance et al., 1981b]. However, if another independent measurement of the distribution were known, such as average energy deposited, this additional constraint should allow a distribution to be constructed. Average energy deposited measurements at a particular fluence measures the energy-weighted distribution function as shown below:

$$<E|\Phi> = \int_0^\infty E \ P(E;\Phi)dE \tag{3.8}$$

In principle absorption measurements and yields, together, should allow the shape of the energy distribution function to be characterized. Black et al. (1979) used this procedure to determine the energy distribution in SF_6, which is thermal and can be represented by a temperature. Similar studies on ethyl acetate showed that its energy distribution is thermal [Danen et al., 1982]. Temporal behavior of the population distribution cannot be obtained from these models, hence dynamical properties of the IRMPE process are ignored.

An alternate approach to extracting information about time evolving populations in quasicontinuum and dissociative continuum levels is from a master equation treatment utilizing bulk measured observables. As shown in section 2, such a treatment requires knowledge of a number of fundamental spectroscopic and kinetic parameters for the reactant molecule. Once all the necessary input parameters are available, populations can be calculated from numerical solutions of the master equations. From the resulting populations, macroscopic behavior of the ensemble of molecules can be predicted by averaging over the populations. Populations predicted by a master equation treatment for chloroethane are shown in Figure 3.1. The calculations are performed with a 80 nsec laser pulse at 968 cm^{-1} for 10 Torr chloroethane at 300 K. At very low fluences only a small fraction of the molecules are excited, and those into low levels. As the fluence is increased, a greater fraction of the population is excited, but still to vibrational levels below 1 eV. At still greater fluences, the distribution is pushed to higher energy and broadens considerably. At the highest fluence in these calculations, population "relaxes" out of the high-energy tail of the distribution as molecules are lost by dissociation. To establish whether these calculated populations agree with experimental fact, direct measurements of $P(E,t)$ are required. As we have seen, distribution-averaged measurements of bulk properties are incapable of giving this information at the necessary level of detail. State-specific spectroscopic probes of level populations are therefore required; a number of experimental approaches to this problem are described in the following section.

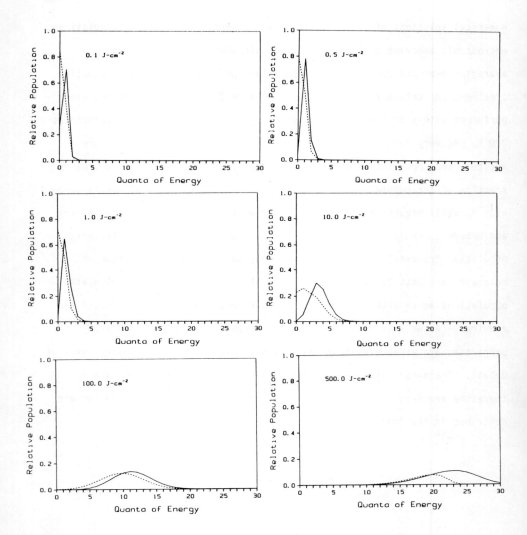

Figure 3.1. Result of a master equation calculation for 10 Torr of chloroethane at several fluences. The solid line shows the population distribution immediately following the laser pulse; the dashed line is the distribution 100 nsec later. [From E. Schweitzer, Ph.D. Thesis, Massachusetts Institute of Technology, © 1985.]

B. State-specific spectroscopic techniques

In order to assess the validity of various IRMPE models, we must have information on the behavior of the level populations in time, and how this behavior depends on parameters such as laser fluence and intensity. In principle, this information can be obtained by carrying out time-resolved spectroscopic measurements on a system undergoing IRMPE. The various techniques available for doing this have been reviewed [Steinfeld, 1983 a,b], and are shown schematically in Figure 3.2. Transient absorption techniques can be carried out in the infrared (IRDR) or optical (IR-UVDR) regions of the spectrum; when a suitable transition is available in the molecule, laser-excited fluorescence excitation and emission spectroscopy (LIFE^{2}S), with its high sensitivity, energy resolution, and time resolution, is to be preferred. Raman probes have also been attempted, including nonlinear techniques such as Coherent Anti-Stokes Raman spectroscopy (CARS). In this section, we shall describe each of these techniques briefly, indicating the type of results that can be obtained, as well as the inherent limitations of each technique.

i. Infrared double-resonance and fluorescence

One of the earliest spectroscopic techniques to be employed was infrared absorption, using a second c.w. CO_2 laser as the probe source for molecules pumped by a high-power pulsed CO_2 laser. This "double-resonance" technique was first applied in our laboratory to SF_6 [Burak et al., 1969; Steinfeld et al., 1970], using a Q-switched CO_2 laser to excite the $v_3=1$ level. Subsequently, $BC\ell_3$ [Houston, 1973], C_2H_4 [Yuan et al., 1973], and CH_3F [Preses and Flynn, 1977] were studied in this way. These experiments provide information on vibrational relaxation among the lower levels of the molecule, but in general do not possess either the frequency or time resolution to explore finer details of the excited-state spectroscopy or inter-mode energy transfer.

124

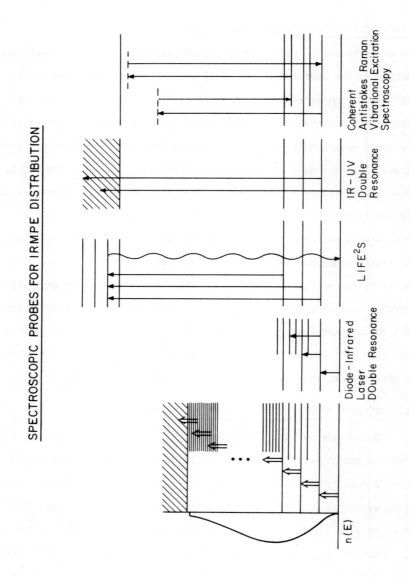

Figure 3.2. Survey of various double-resonance probing techniques. The first column represents the IR pumping of the molecule. Infrared and ultraviolet absorption double resonance are shown in columns 2 and 4, respectively. Column 3 is laser-excited fluorescence, and Column 5 portrays a typical Raman technique, namely, CARS [Steinfeld, 1983a].

By using a high-power TEA-CO_2 laser as the pump source, the molecules can be raised to considerably higher excitation levels: 5 to 10 CO_2 laser quanta, which is well into the quasicontinuum. Such measurements have been performed, e.g., for SF_6 [Frankel, 1976; Fuss, 1978, 1980; Lyman et al., 1980; Bagratashvili et al., 1979]. The problem with these measurements is that the spectra that are obtained tend to be completely different from experiment to experiment. This results from the energy distributions depending on details of the pulsed laser mode structure, SF_6 pressure and temperature, and other parameters to an extent that makes such measurements very difficult to interpret quantitatively.

Infrared fluorescence is occasionally used as an alternative probe technique [Glatt and Yogev, 1981]. The spectroscopic bandwidth in IR emission experiments is generally too broad to permit resolution of individual states, although the "red shift" which is frequently observed in the spectrum is interpreted as emission from high-lying vibrational levels. Furthermore, the IR emission lifetime is so long that the molecules may undergo considerable collisional relaxation prior to emission. An exception is the work of Frankel and Manuccia [1978], who used a mode-locked CO_2 laser having a 3 nsec pulse width to excite SF_6 in the v_3 manifold. Fluorescence was observed at 16 μm, corresponding to $\Delta v_4 = -1$ transitions, with a rise time much too fast to be attributable to a collision-induced V-V process. From this work, the authors concluded that rapid intramolecular vibrational relaxation must be taking place. Similar observations were subsequently made by Hudgens and McDonald [1985], who observed the v_3 fluorescence from laser-pumped SF_6 in the 850-950 cm^{-1} region, and interpreted their results in terms of a Boltzmann distribution in the vibrational population. In this connection, the work of McDonald and co-workers should be mentioned [Stewart et al., 1984; Kulp et al., 1984, 1985]; by using a supersonic beam of molecules and a liquid-nitrogen-cooled Fourier Transform spectrometer, they are able to overcome many of the limitations on resolution and collisional relaxation inherent in conventional infrared fluorescence techniques.

ii. Diode-Infrared Laser-DOuble Resonance

The use of fixed-frequency sources for double resonance imposes a severe limitation on the technique. The CO_2 laser, for example, has spacings of ~ 1.5 cm^{-1} between lines, so if the width of individual spectral features is on the order of 40-50 MHz, only 1/1000th of the spectrum can be sampled. This limitation may be removed by the use of a tunable infrared probe source; the lead-salt diode laser is particularly suitable for this, since it provides narrow-bandwidth (<0.001 cm^{-1}) radiation in the CO_2 laser wavelength bands. The first system to be investigated by this technique was SF_6 [Jensen et al., 1979; Moulton et al., 1977; Reiser et al., 1981; Dubs et al., 1982]. These experiments provided the first detailed rotational energy-transfer measurements in a spherical-top molecule; in addition, information on the $2v_3 \leftarrow v_3$ transition was obtained [Dubs et al., 1982; Patterson et al., 1981]. Other molecules which have been studied in this way include methane ($^{13}CD_4$) [Laux et al., 1984] and fluoroform (CDF_3) [Harradine et al., 1984]; the latter experiments will be discussed in a little greater detail as an illustration of the technique.

The apparatus used in these experiments is depicted in Fig. 3.3. A Q-switched CO_2 laser (or other pulsed IR source) is used as the pump; that radiation propagates through a low-pressure gas sample and is overlapped by the diode laser probe beam. The probe beam is detected by a fast, cooled IR detector and is demodulated by a lock-in amplifier, gated integrator, or transient recorder, depending on whether one is measuring the normal transmission spectrum, the induced double-resonance spectrum, or the time

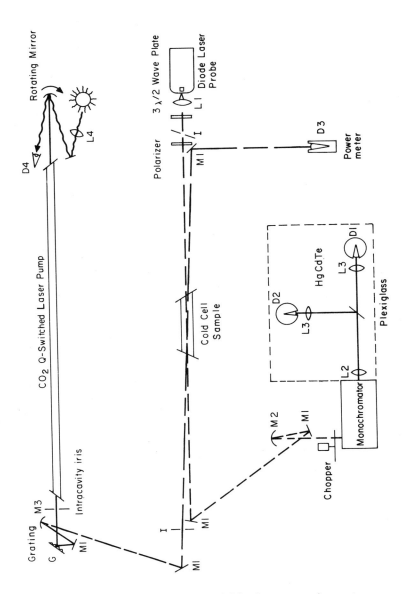

Figure 3.3. Typical Diode-Infrared Laser-DOuble Resonance Apparatus [Harradine et al., 1984].

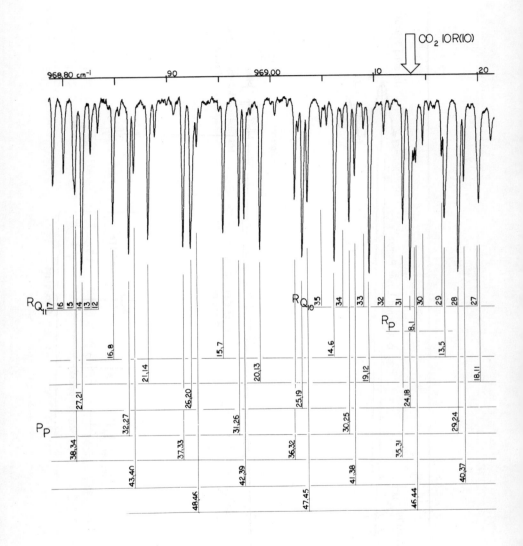

Figure 3.4. Absorption spectrum of CDF_3 at 173 K in the vicinity of the CO_2 10R(10) laser line at 969.1395 cm-1. [Harradine et al., 1984].

dependence of the signal at a fixed frequency, respectively.

The basis for the state-selective pumping is shown in Fig. 3.4, which displays a small portion of the IR absorption spectrum of CDF_3. The CO_2 10R(10) laser line coincides most closely with the $^RP_1(8)$, $^PP_{44}(46)$, and $^PP_{18}(24)$ transitions of $v_5\leftarrow 0$, which populate (J'=7, K'=2), (J'=45, K'=43), and (J'=23, K'=17) of the $v_5=1$ vibrational level, respectively. The double-resonance scheme expected for the latter transition is shown in Fig. 3.5. Three-level signals from the depleted lower level appear at the $^PQ_{18}(24)$, $^PR_{18}(24)$, $^RP_{18}(24)$, $^RQ_{18}(24)$, and $^RR_{18}(24)$ transitions of $v_5\leftarrow 0$ (a two-level signal can be observed at the pumped $^PP_{18}(24)$ line, but would show considerable power broadening [Jensen et al., 1979]). New transitions appear as excited-state absorptions (esa's) at the $^PP_{17}(23)$, $^PQ_{17}(27)$, and $^PR_{17}(23)$ lines of the $2v_5(\ell=2)\leftarrow v_5(\ell=1)$ band, and at $^RP17(23)$, $^RQ17(23)$, and $^RR17(23)$ of the $2v_5(\ell=0)\leftarrow v_5(\ell=1)$ component. In addition, as population is transferred to or from the pumped levels by collisions, additional signals (four-level signals) appear at neighboring lines; the collisional processes giving rise to these signals are indicated by wavy lines in Fig. 3.5.

Several of these transitions are shown in Fig. 3.6. The $^RQ_{18}(24)$ transition appears as a strong negative-absorption signal coincident with the pump. This signal decays very rapidly due to rotational energy transfer in the molecule, and additional four-level signals appear at neighboring frequencies in the $^RQ_{18}(J)$ band. An extended scan of the spectrum shows that, in addition to this three-level signal, esa's appear at the $^RQ_{17}(23)$ and $^PQ_{17}(23)$ frequencies predicted in Fig. 3.5. The additional features labeled $^PQ_{19}$ and $^RQ_{16}$ cannot be connected with the J'=23, K'=17 level, because $\Delta K=\pm 3n$ is a strong propensity rule in rotationally inelastic collisions of CDF_3. Instead, these arise from relaxation out of the J'=45, K'=43 level which is

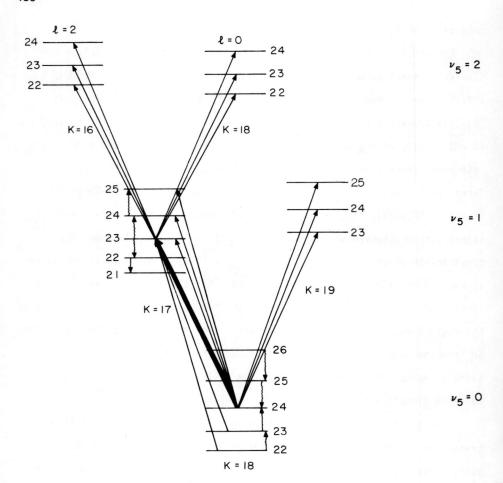

Figure 3.5. Energy level diagram of all possible three-level double-resonance signals resulting from pumping CDF_3 with the $10R(10)$ CO_2 laser line, indicated by the heavy arrow [Harradine et al., 1984].

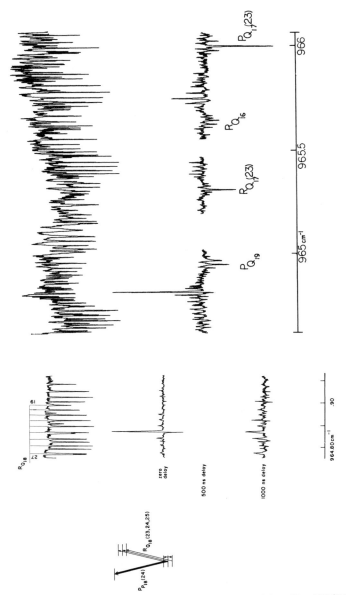

Figure 3.7. IR double-resonance signals in CDF₃ pumped by CO₂ 10R(10). The left-hand spectrum (with accompanying level diagram) is for 70 mTorr CDF₃ cooled to 185K. In the extended scan (right), 1 Torr CH₂F₂ is added to enhance rotational relaxation [Harradine et al., 1984].

also pumped by the CO_2 10R(10) line. This and other evidence shows that K can change by large values in a collison. The energy transfer pathways revealed by these experiments play an important rôle in overcoming the rotational bottleneck to IRMPE in this molecule [McAlpine et al., 1983]. The use of higher-power CO_2 or other pump lasers will furnish additional information on excited-state spectroscopy and IRMPE dynamics.

iii. Infrared-ultraviolet double resonance.

In principle, probing IRMPE at visible or u.v. wavelengths should offer advantages over probing at infrared wavelengths. In the infrared, absorptions from vibrationally excited origins are displaced from the fundamental by only small anharmonicity terms, leading to overlapping bands; but in electronic spectra, different harmonic constants in the ground and excited states can lead to long, well-resolved vibrational progressions. Also, the much higher brightness of tunable u.v. and visible laser sources as compared with diode lasers, should facilitate detection and signal averaging. When fluorescence spectroscopy is feasible, it can indeed be an excellent probe of IRMPE (following section); but even if the molecule of interest does not possess a fluorescing state at an accessible energy, information can still be derived from (u.v./visible)-i.r. absorption double resonance.

The ozone molecule [Adler-Golden and Steinfeld, 1980; Adler-Golden et al., 1982] affords a good example of this technique. Ozone possesses a strong u.v. absorption (Hartley band) in the 210-300 nm region, and its v_3 (asymmetric stretch) mode coincides with several CO_2 laser lines in the 9.6 µm band. By measuring the u.v. absorption transient, the extinction coefficient for the vibrationally excited level could be obtained, and related to a potential surface model for the upper electronic state. From the decay times of these transients, rate coefficients for V-V and V-T energy transfer could be obtained. CF_3I has also been studied in this way, using either flash kinetic spectroscopy [Padrick et al., 1980; Fuss, 1982] or absorption of u.v. laser lines [Kudriavtsev and Letokhov, 1980]. The "red shift" of u.v. absorption spectra commonly observed in these experiments was proposed as the basis for a molecular laser isotope separation process using UF_6; the pertinent spectra have recently been published [Kim et al., 1985].

iv. Laser Induced Fluorescence Excitation/Emission Spectroscopy

When a molecule does possess a fluorescent electronic transition at a wavelength accessible to tunable u.v./visible lasers, LIF can provide a great deal of information on infrared pumping and rotational dynamics. An excellent example is afforded by the work of the New South Wales group on formaldehyde, particularly D_2CO. In addition to providing improved spectroscopic resolution of the u.v. absorption bands [Orr and Nutt, 1980; Orr and Haub, 1984], a considerable amount of information was obtained on rotational [Orr et al., 1981, 1984] and vibrational [Haub and Orr, 1984; Bewick et al., 1985] relaxation among laser-pumped levels of the molecule. The finding that large collision-induced K changes take place in D_2CO is analogous to the results mentioned earlier for fluoroform [Harradine et al., 1984]. Another significant finding is that the $v_4=1$ and $v_6=1$ vibrational levels are strongly coupled by Coriolis interactions, leading to rotationally selective vibrational energy transfer between these modes. The implication for IRMPE is that highly efficient rotational redistribution and mode-mode transfer, both due to collisions, play an important rôle in overcoming the bottlenecks in the lower levels [Haub, 1985].

A similar experiment on benzene has recently been reported [Lyman et al., 1985]. Because of the considerably higher vibrational and rotational congestion in this molecule, as compared with formaldehyde, the results had to be interpreted with the use of a theoretical model. It is somewhat surprising that, even in this "large" molecule, collisions are required to facilitate infrared pumping. Related experiments have also been carried out on propynal [Brenner et al., 1980].

The measurements cited above all probe the first one or two excited levels in the vibrational pumping sequence. Attempts to probe the multiple-photon excitation regime have proven to be difficult. Experiments on thiophosogene [Brenner, 1981; Brenner et al., 1981, 1983] have shown that, when the $2v_4$ mode is pumped by a CO_2 laser, population is efficiently removed from levels $v_4 = 0,1,2,3,4,6$, and 8, but does not appear in higher levels or in other modes. This effect is operative over a wide range of laser frequencies, as shown in Fig. 3.7, which does not correlate at all with the small-signal infrared absorption coefficient of thiophosgene. The pumping efficiency is "quenched", rather than enhanced, by added buffer gases at a rate much faster than gas-kinetic collision frequency. These observations are not fully understood at the present time.

A related phenomenon occurs when an electronically excited state of a polyatomic molecule can absorb CO_2 laser radiation, while the ground-state can do so only weakly or not at all. In this case, IRMPE in the electronically excited state can lead to a shift in wavelength and decrease in lifetime of luminescence, when the triplet state is vibrationally excited to an energy at which it can mix with the lowest excited singlet state. This occurs in biacetyl [Borisevich et al., 1979; Burak et al., 1979b; Tsao et al., 1980, 1982] and propynal [Stafast et al., 1981]. In the case of NO_2 [Feldmann et al., 1980], IR pumping of the $^2B_2$ state both enhances the visible fluorescence and leads to dissociation into NO + O.

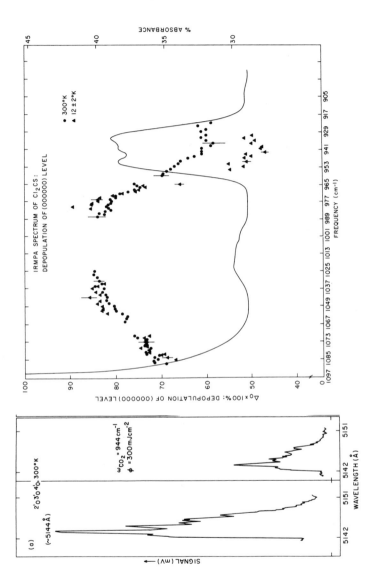

Figure 3.7. LIF monitoring of IR pumping in thiophosgene. Left: excitation spectra of $2_0^1 3_0^1 4_0^1$ band with and without infrared pumping, showing considerable ground-state depletion. Right: dependence of depletion effect on CO_2 laser line frequency, as compared with i.r. absorbance of thiophosgene [Brenner et al., 1983].

v.   Infrared-Raman Double Resonance

Many molecular systems in which IRMPE is an important process do not possess electronic transitions in a wavelength range convenient for monitoring by either LIF or double-resonance absorption. In principle, this limitation can be circumvented by the use of Raman spectroscopy as a probe of vibrational populations. Raman scattering does not require the presence of a resonant transition and, if carried out with pulsed, tunable u.v. or visible laser sources, can possess state specificity and time resolution comparable with LIF. Its major drawback is the very small Raman cross-section, which severely limits sensitivity. This can be partially overcome by the use of resonance Raman (but if a resonant state is available, LIF is usually preferable) or coherent and/or stimulated processes such as Coherent Anti-Stokes Raman Scattering (CARS) or Stimulated Raman Gain (SRG).

IR-Raman experiments have been carried out on SF_6 and CF_3I by Bagratashvili et al. [1979, 1980, 1981]. The weak Raman signals made analysis difficult, but evidence for a "two-ensemble" distribution was obtained. This may have been due to the Raman probe sampling molecules both in and out of the IR laser field. In an IR-CARS double resonance experiment [Alimpiev, 1982], structure was observed at the long-wavelength side of the Raman-allowed $v_1(A_{1g})$ band which was assigned to origins in n vibrational quanta of the pumped $v_3(F_{1u})$ mode. In subsequent experiments by Schweitzer [1985] on SF_6 and other molecules, these features were not observed (see Fig. 3.8). Indeed, the non-appearance of population in upper vibrational levels as the ground state is depleted by i.r. pumping is very reminiscent of the behavior seen in thiophosgene [Brenner et al., 1981, 1983]. Results of a time-resolved Raman study of SF_6 [Mazur et al., 1984] were interpreted as indicating collisionless IVR, but this interpretation is not unambiguous. The SRG technique has also been applied to laser-pumped SF_6 [Esherick et al., 1982]; rotational resolution comparable with diode laser IRDR was obtained, but only for the lowest excited vibrational level.

vi. Other spectroscopic techniques

Several additional probe techniques have been employed in connection with IRMPE. Photoionization with mass-spectrometric detection has been used to monitor dissociation products in the IRMPD of acetic anhydride [Feldmann et al., 1983], but this technique has not been used to probe for vibrational origins in the parent molecule following IRMPE. Stimulated-emission pumping [Hamilton et al., 1986] provides an alternative technique for preparing molecules in highly excited vibrational states. This technique has been applied to several systems, most notably formaldehyde, acetylene, and p-difluorobenzene, but is restricted by the requirement of a resonant transition at a u.v. or visible wavelength suitable for optical pumping.

Highly excited vibrational levels can also be prepared by photoexciting to an electronically excited state which undergoes rapid internal conversion. Molecules studied in this way include azulene [Rossi et al., 1983; Hippler et al., 1985], toluene [Hippler et al., 1983a; Heymann et al., 1984], cycloheptatriene [Hippler et al., 1983b] and CS_2 [Dove et al., 1985]. These measurements have provided important information on vibrational deactivation processes for high vibrational levels, which is an important component of IRMPE models. At the other extreme, IR excitation has been used to dissociate molecular clusters formed in supersonic molecular beams. For these systems, absorption of one photon is usually sufficient to break the weak van der Waals bond. Space limitations preclude a more extensive discussion of either of these topics in the present review.

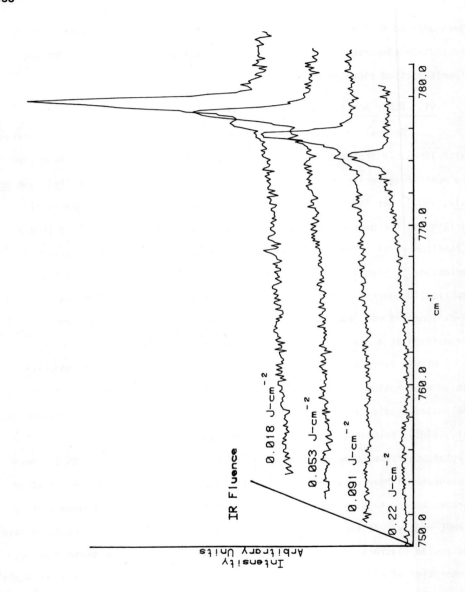

Figure 3.8 CARS spectra of 2 Torr SF_6 at 200 K pumped with CO_2 10P(20) laser line at 100 nsec delay for a range of i.r. fluences (0.018 - 0.22 J/cm^2) [from Schweitzer, 1985].

vii. Summary of properties derived from state-specific measurements

As compared with the highly averaged bulk measurements, spectroscopic probes can provide time-resolved distribution functions over specific quantum states, at least for the lower vibrational levels of molecules. One result which has emerged from both diode laser IRDR and IR/UV LIF measurements is the importance of highly efficient, rotational state-specific V-V transfer between Coriolis-coupled vibrational modes in overcoming the "bottleneck" in the initial steps of IRMPE. Detailed information on higher vibrational levels has proven to be much more elusive, however.

The absence of spectra attributable to specific excited vibrational origins in IRMP-excited small molecules, in both LIF [Brenner et al., 1981, 1983] and CARS experiments [Schweitzer, 1985], suggests that the molecules which are pumped by the laser are distributed rapidly over a wide distribution of vibrational levels. This would be consistent with IRMPE models invoking rapid IVR, rather than those in which a pumped mode persists well into the quasicontinuum. The major problem in interpreting spectroscopic probes at high excitation levels has been clearly stated in the literature [Bagratashvili et al., 1979; Steinfeld, 1983b], namely, that neither the absorption spectrum $\sigma(v)$ or the distribution of absorbers $N(v_i, J_i)$ are well known. The convolution of this dual lack of knowledge makes any unique assignment of transient spectra which may be observed quite difficult. Nevertheless, both pieces of information are of considerable interest, and should continue to be pursued.

4. APPLICATIONS OF INFRARED MULTIPLE-PHOTON EXCITATION

A. Application of IRMPD to chemical kinetics and synthesis

A complete understanding of the time evolution of complex chemical systems can be achieved only with knowledge of individual elementary reaction steps. Often, these elementary steps involve reactive intermediates which are difficult to produce and monitor. With ultrasensitive detection methods employing lasers, the tools are available for monitoring reactive intermediates; but the study of these intermediates is limited by the need to generate these species cleanly. A large number of molecular dissociation and isomerization reactions are induced by IRMPE; indeed, it appears that any molecule with more than three or four atoms, which possesses an absorption band at or near a strong laser output wavelength, can be made to undergo this process. Since the products are generally specific molecular fragments, the temperature of the reaction system can be varied independently of the excitation process, and side products resulting from either wall or secondary gas-phase reactions can be reduced or nearly eliminated, IRMPD is an extremely attractive method for generating reactive intermediates for kinetic studies.

i. Generation of Reactive Intermediates by IRMPD

Thermal reaction of radical precursors in heated vessels has long been the primary source for generating reactive intermediates. Interpretation of kinetic data is often obscured by secondary processes which occur on the walls of the reaction vessel. Secondary wall reactions are minimized by the use of low-power light sources such as resonance lamps or broadband sources in conventional photolytic methods, but the method produces low number densities of reactive intermediates. The use of visible and u.v. laser sources can give rise to excited reactive intermediates in high concentration; however, u.v. photodissociation may give rise to a distribution of products. IRMPD has several advantages over the above sources. Since molecules which undergo IRMPD react through the lowest energy channel, and the thermal chemistry is well known for most molecules, reactive intermediates resulting from dissociation can easily be predicted. Furthermore, the intermediate can be produced directly, homogeneously and in high concentrations in the laser beam free from undesired catalytic effects at the vessel surface, which have always plagued thermal experiments.

The success of generating reactive intermediates depends on whether a suitable precursor for the intermediate can be found. The choice of precursor involves several considerations. It must have strong absorption bands which overlap with available laser lines. To produce intermediates from a bond fission process, it would be advantageous to choose a precursor with the lowest bond strength since this leads to a low threshold for bond breaking. This results not only in the production of homogeneous intermediates at low fluences, but also avoids tight focusing conditions of the laser beam which could introduce undesirable side products. Table 3.1 gives a brief list of some kinetically important intermediates produced by IRMPD using CO_2 lasers. Care must be taken, however, to avoid secondary dissociation of the newly formed intermediate during the laser pulse, which would produce unwanted products at the expense of the primary intermediate. Some chemical systems in which secondary IRMPD has been reported include fluorohalocarbons [Würzberg et al., 1978; Rossi et al., 1982; Horwitz et al., 1981] and alkylsilanes [Francisco et al., 1984]. Usually, secondary photolysis can be minimized by careful choice of excitation wavelength, operating laser power, and laser pulse duration. Small radicals are less susceptible to further excitation, and dissociation of the primary species can be suppressed. If conditions are optimized, high yields of the primary intermediate can be obtained

[Bagratashvili et al., 1984].

An illustration of these principles is the CF_3 radical, which is important in both atmospheric photochemistry and semiconductor surface chemistry. CF_3I is a suitable precursor for this radical, since it has a strong absorption coincident with the 9.6 μm region of the CO_2 laser and the C-I bond strength is very low (ca. 52.6 ± 1.1 kcal mol^{-1} [Okafa and Whittle, 1973]). Furthermore, IRMPD studies demonstrate that CF_3 radicals are produced in copious amounts at low laser fluences [Bittenson and Houston, 1977; Wurzberg et al., 1978]. We have used the IRMPD/LIF method to study secondary dissociation of CF_3 produced from CF_3I, CF_3Br, and C_2F_6 precursors [Thoman et al., 1985]. In each case, CF_2 is detected (via its $\tilde{A}^1A_1 \leftrightarrow \tilde{X}^1A_1$ transitions), but only in low yield and at fluences considerably above the primary dissociation threshold.

Secondary photolysis can also be circumvented by delayed production of the primary reactive intermediate; this is achieved by generating the species by secondary reaction after the laser pulse. This method was developed by Slagle and Gutman (1982). They used IRMPD to generate chlorine atoms which react with organic precursors in an atom-molecule abstraction reaction, which then produces the desired intermediate. Acetyl radicals, which could not be generated directly by IRMPD because a suitable precursor was unavailable, were generated using this method by the following scheme:

$$CF_2C\ell_2 \xrightarrow{nh\nu} CF_2C\ell + C\ell \tag{4.1a}$$

$$C\ell + CH_3CHO \longrightarrow CH_3\overset{\centerdot}{C}O + HC\ell \tag{4.1b}$$

The authors observed no significant photolysis of the acetyl radicals and found them to be produced in stoichiometric yields provided high precursor concentrations were used.

ii. Utilization of IRMPE for kinetic measurements

Several new techniques have adapted IRMPE and IRMPD to familiar
kinetic methods for studying reaction mechanisms and rates in real time. The
principal innovative feature that is contributed by these modified techniques
is the capability of producing reactants under clean, well-defined conditions.
This, coupled with sensitive detection techniques to monitor the time
evolution of the intermediates, has allowed individual elementary reaction
steps to be interrogated and is resulting in accurate rate constant
measurements for many species of importance to greater understanding of
atmospheric, combustion, and surface chemical processes.

One such method is the infrared-laser photolysis/mass spectrometry method
which includes the very low-pressure photolysis (VLPΦ) technique. These
methods operate on the principle that a reactive gas continuously flowing
through a photolysis cell is photodissociated by IRMPD, and the resulting
reactants and products are detected in real time by a mass spectrometer
directly coupled to the cell. Martinez et al. [1980] have used this technique
to study the recombination reaction of CF_2 radicals produced by IRMPD at
high pressures (1.8 - 2.7 Torr) of $CF_2HC\ell$. The VLPΦ technique, which operates
under the same principle, has been used to study radical-molecule bimolecular
reactions. Reactions using the VLPΦ technique are studied at low pressures
(<1 mTorr) of radical precursors, which insures that bimolecular collision
processes occur only with the added nonabsorbing reagent gas. This technique,
developed at SRI, has been used to measure absolute thermal rate constants for
CF_3 with NO and O_3 over a wide temperature range [Rossi et al., 1979]. Yamada
et al. (1981) have also monitored the reaction of CH_3 radicals with NO_2 using
photoionization mass spectrometry; the CH_3 radical was produced by IRMPD of
$C_6F_5OCH_3$.

Another technique which is proving to be of great utility in kinetic

measurements is the IRMPD photolysis/chemiluminescence method. It employs
IRMPD to initiate reaction from an atom or radical precursor. If the reactive
intermediates produced from the reaction are formed in vibrationally or
electronically states which emit strongly, they can be followed during the
course of reaction by the time resolved detection of that emission; reaction
rate constants are measured from the appearance rate of the emitted
fluorescence. Sometimes reactive intermediates are formed in the ground state
or in nonemitting states; in this case, species can be monitored by
chemiluminescence resulting from its reaction with another reagent. The
development of these techniques for rate constant determinations has been
explored by Houston and co-workers [Wurzberg et al., 1978; Wurzberg and
Houston, 1980; Grimley and Houston, 1980; Houston, 1981]. The technique has
been used to study a number of reactions. Wurzberg and Houston (1980) studied
hydrogen abstraction by fluorine atom using SF_6 as the atom precursor; rates
of the $F+H_2(D_2)$ reaction were determined from monitoring $HF^{\ddagger}$ and $DF^{\ddagger}$
chemiluminescence over a range of pressures (120-400 mTorr) and temperatures
(190 to 373°K). Reactions of more elusive radicals have been investigated
using this technique, for example the reaction of $C_2(a^3\Pi_u)$ with O_2 (Filseth et
al., 1979) and NO (Levy et al., 1980). In both cases the C_2 radical was
generated by IRMPD of C_2H_3CN. Another example is the photochemical reaction
of $ND(a^1\Delta)$ with DN_3 which proceeds according to the scheme [Hartford, 1978]

$$ND(a^1\Delta) + DN_3 \longrightarrow ND_2(^2A_1) + N_3 \qquad (4.2)$$

$ND(a^1\Delta)$ was prepared by IRMPD of DN_3, and the fluorescence from $ND_2(^2A_1)$
permitted rate constant measurement. Because of the wide variety of
molecules which can undergo IRMPD, the photolysis/chemiluminescence method is
proving to be a very useful technique for interrogating individual elementary
reaction steps in complex reaction mechanisms involving reactive

intermediates, atoms, and stable molecules.

The IRMPD/laser-induced fluorescence (LIF) method is another technique currently used to study kinetics of important reactive intermediates produced by IRMPD. Detection of intermediates by LIF requires that the intermediate possess a bound excited electronic state from which fluorescence can be induced by tuning the laser wavelength to the appropriate electronic transition of the intermediate. Since the only molecules excited are those which correspond to the specific transition in the intermediate, the excitation is selective and allows unambiguous identification of intermediates. Furthermore, the LIF technique can be extremely sensitive; molecular densities in the range of 10^4 -10^5 cm^{-3} have been successfully detected. Lesiecki et al. [1980] have studied reactions of C_3, produced in the $^1\Sigma_g^+$ state, by IRMPD of allene (C_3H_4) with NO and O_2 using LIF with a dye laser probing the $^1\Pi_u \leftarrow {}^1\Sigma_g^+$ transition of C_3.

The methods mentioned above rely on direct IRMPE of reactive intermediate precursors. Direct excitation is not possible for some chemical systems due to absence of a strong absorption band coinciding with IR laser wavelengths. Reaction can still be induced by a sensitizer, which is a nonreacting molecule that absorbs the IR energy. The energy deposited in the sensitizer molecule is transferred to the reactant molecule, not resonant with the laser, via collisions during and after the pulse, thereby establishing thermal equilibrium. Consequently, IRMPE can be used as a means of initiating reactions through thermal heating without surface heating, hence suppressing heterogeneous reactions occuring at the gas-surface interface, which could complicate kinetic analyses through the initiation of long chain reactions. The most commonly used sensitizer is SiF_4. The choice is based largely on its high absorptivity, stability, and lack of reactivity in the presence of a wide variety of substances. The rapid heating which ensues can be used in a temperature-jump experiment, as in the work on the $2NO_2 \leftrightarrow N_2O_4$ equilibrium by Gozel et al. [1984]. Another molecule also used is SF_6, as in the pioneering experiments on sensitized laser pyrolysis by Shaub and Bauer [1975]. Laser sensitized pyrolysis, sometimes referred to as laser-powered homogeneous pyrolysis (LPHP), has been applied to a large number of organic synthesis and mechanism studies. These have been reviewed by Danen and Jang [1981]. McMillen et al. [1982] and Dai et al. [1982] developed LPHP as a general tool for measuring rate constants. Since these studies LPHP has proven to be a significant technique for studies of homogeneous reactive processes.

This is illustrated by the use of LPHP in resolving a long standing controversy over the mechanism of silane pyrolysis. The primary dissociation step in silane decomposition proceeds through a three-center elimination reaction to form silylene and molecular hydrogen, as shown in static pyrolysis [Purnell and Walsh, 1966; Neudorf et al., 1980] and shock tube [Newman et al., 1979] studies. These works all assert that the reaction

$$SiH_4 \longrightarrow H_2Si\overset{\text{H}}{\underset{\text{H}}{\vdots}} \longrightarrow :SiH_2 + H_2 \tag{4.3}$$

is homogeneous. However, results of Robertson et al., [1984] for silane pyrolysis under static and flow conditions suggest that the initiation mechanism is exclusively heterogeneous; this led the authors to claim that all previous studies of silane pyrolysis involved heterogeneous processes. Jasinski and Ester (1985) helped resolve the mechanism issue using the LPHP technique. In their study, cyclopropane is used as an internal standard to monitor the temperature history of the reaction occuring in the irradiation volume of the laser. The cyclopropane undergoes unimolecular isomerization to propene; simultaneously, the SiH_2 produced in (4.3) inserts into undissociated SiH_4 to form Si_2H_6; H_2 and a powdery deposit of amorphous silicon are by-products of the latter reaction. From relative yields of propene and Si_2H_6, rate information for each reaction channal can be deduced. By measuring the relative rates as a function of varying cell conditions (clean or coated with powder from previous experiments); reactant, sensitizer, and internal standard pressures; various cell designs; added free radical scavenger; and laser conditions, Jasinski and Ester could not find systematic changes in the relative rates. These results suggested that the unimolecular reaction of silane in static systems is indeed homogeneous.

iii. Mode selective chemistry

From the large number of kinetic studies which have utilized IRMPE and IRMPD, it appears that the chemistry induced by the laser is no different from the thermal chemistry. Indeed, the term "laser chemistry", which is occasionally encountered, is misleading; while there is photochemistry induced by lasers, there does not appear to be any chemistry unique to laser excitation. One "holy grail" which has frequently been mentioned is the use of lasers to selectively excite and break a particular bond in a molecule, or to cause reaction to occur only at that site. Much of the initial impetus of IRMPE studies was to search for chemical systems which exhibited potential mode selective chemistry. Encouragement was furnished by the work of Dever and Grunwald [1976] which claimed that energy in $CFCl_3$ and CF_3Cl remained localized in a single vibrational mode of these molecules following IRMPE long enough for reaction to occur, based on the observed reaction products. Claims of mode-selective decomposition of octafluorocyclobutane had been made using similar arguements by Preses et al. [1977]. These conclusions have been refuted by Schulz et al. [1979] and Cantrell et al. [1979], who have given alternate interpretations of the data without invoking energy localization claims. Danen and Jang [1981] have reviewed claims and counter claims of mode selective reaction for numerous molecules appearing in the literature. They concluded that there is no clear-cut evidence for mode-specific chemistry in any of the systems studied.

The possibility of mode selective chemistry relies on the competition of vibrational excitation in a particular mode or restricted set of modes with rapid intramolecular redistribution of the deposited laser energy. As shown in Table 2.1, if the IVR process is faster than the laser pumping rate, the energy deposited in the molecule randomizes quickly as it is absorbed, hence no selective chemistry can be expected. If, however, IVR is slower than the laser pumping rate and the dissociation rate, then selectivity may be possible. The critical consideration is thus whether there is any possibility of exciting a molecule in such a way that the initial energy remains localized on a time scale long compared with reaction. The use of picosecond infrared excitation has been proposed, but the relevant time scale is still that of the reaction, rather than the excitation. Another possibility may be enhancement of bimolecular reactions possessing a low activation energy; in this case, the vibrational energy content of the molecule may be kept low enough to slow down IVR, but still be sufficient to promote the reaction in collision with a second species. To date, however, the only clear evidence for such mode-selective rate enhancement is that of Pimentel and coworkers [Frei and Pimentel, 1983, 1985; Knudsen and Pimentel, 1983]. In this work, the ethylene-fluorine (C_2H_4/F_2) and allene-fluorine (C_3H_4/F_2) systems were investigated in low-temperature matrices, and certain modes were found to enhance the reaction in excess of energy content alone. This suggests that mode selectivity may be possible only in low-activation-energy reactions which are artificially slowed down by matrix constraints and low temperatures;

nevertheless, definition of the conditions under which mode selectivity may occur is a valuable contribution to our understanding.

iv. Limitations on synthetic utility

Despite the clear usefulness of IRMPE in preparing reactive intermediates for chemical-kinetic studies, no commercial process for production of chemicals by IRMPE has yet made its appearance. The principal reason for this is one of economic limitations, rather than technical feasibility (although a proposed process for enrichment of uranium by isotope-selective IRMPE of UF_6 was abandoned in part because of the lack of a suitable laser at the UF_6 16μm absorption wavelength). Several examples will illustrate this limitation.

Hydrazine (N_2H_4) can be produced cleanly and in reasonable yield by IRMPE of methylamine (CH_3NH_2), without formation of undesirable side products such as nitrosamines [Hwang and Sandstrom, 1981]. The energy cost of running the laser to produce this material works out to be \$2300/lb, which is three orders of magnitude <u>greater</u> than the current selling price of hydrazine! Not quite as unfavorable, but still uneconomical, is the synthesis of COF_2 reported by Malatesta <u>et al.</u> [1984]. In that report, 10 5-Joule CO_2 laser pulses are required to produce on the order of 10^{-4} moles of COF_2. The cost of the laser energy (50J $\simeq 4 \times 10^{-3}$ "moles" of 10-μm laser photons) is about 25¢, based on the figures presented by Reddy [1984]; the value of the COF_2 produced is less than 2¢ [SCM, 1984]! Since costs for materials handling, purification, feed-stocks, etc., must be added to the cost of the laser energy, it seems unlikely that direct synthesis by IRMPE will ever become a commercially viable process.

For this reason, most commercial attention has been directed to preparation of intrinsically high-cost materials, such as enriched isotopes. The status of laser isotope separation schemes based on IRMPE for materials

such as deuterium and tritium or ^{13}C has been extensively reviewed in the
literature [Herman et al., 1984; Hackett et al., 1984]. Many of these
schemes have been found to be not only technically but economically feasible.
The major impediment to their implementation is that the market requirements
for all these isotopes is met or exceeded by currently installed capacity, so
there is little incentive to investing in new laser enrichment facilities.

One way of overcoming this limitation is to use the laser to initiate a
chain reaction; in this way, the effective "quantum yield" of product per
photon can be 10^3-10^4 or more. This has been shown to be effective, for
example, in the use of u.v. excimer laser radiation in the synthesis of vinyl
chloride monomer from dichloroethylene [Wolfrum and Schneider, 1984]. The
major difficulty in commercial implementation of that process at the present
time is the unavailability of excimer lasers having sufficiently high output
power.

The most promising application of laser-driven chemistry to emerge so far
is the use of lasers to carry out specific chemical modifications on a
material which already bears significant processing costs. The best example
of this is laser processing of microelectronic devices, which has attracted a
great deal of interest in the past several years. We consider the use of
IRMPE for such processes in the following section.

B: Surface Chemical Processes Initiated by IRMPE

Using laser initiation, it is possible to produce a high transient
population of reactive intermediates localized in both space and time. This
is particularly the case for IRMPD, since the dissociation takes place
entirely within the focal volume defined in Eq. (2.19). In addition, energy
is deposited only in the volume of material pumped directly by the laser; the
rest of the system can be maintained at (or below) ambient temperature. For

these reasons, laser-initiated reactions appear to be very attractive for carrying out the delicate chemical transformations required in the fabrication of semiconductor microelectronic devices. The economics of such processes are also favorable, since the laser processing represents only a small addition to the total production cost of such devices. A large and increasing number of reports of such processes, using argon-ion, excimer, Nd:YAG, and CO_2 lasers, has appeared in the literature [Bäuerle, 1984; Osgood, 1983; Selwyn and Lin, 1985; Chuang, 1982]. Since the reaction zone cannot be made smaller than the diffraction-limited spot size, which is on the order of a few wavelengths, the use of short-wavelength (visible and u.v.) lasers permits direct writing of surface features with dimensions of 1 μm or less. Also, since the photochemistry at these wavelengths is generally initiated by single-photon absorption, lower powers can be used, thus minimizing surface damage. For the purposes of this review, however, we shall emphasize processes in which IRMPD plays a clear rôle, and may present some distinct advantages.

Such is the case for the reactive etching of silicon-containing materials, which is typically carried out in a fluorocarbon plasma. The mechanism of the reactive etching process is generally supposed to involve the following steps [Winters, 1978; Winters and Coburn, 1979].

dissociation of fluorocarbon in the plasma

$$CF_4 \longrightarrow CF_x + (4 - x)F \qquad (4.4)$$

adsorption of fluorocarbon fragment:

$$CF_x + Si \longrightarrow CF_x(ads) \qquad (4.5)$$

dissociation of adsorbed fluorocarbon fragment:

$$CF_x(ads) \longrightarrow C(ads) + xF(ads) \qquad (4.6)$$

adsorption of fluorine atoms:

$$F + Si \longrightarrow F(ads) \tag{4.7}$$

reaction with surface silicon atoms:

$$yF(ads) \longrightarrow SiF_y (ads) \tag{4.8}$$

desorption of fluorosilyl product:

$$SiF_y(ads) \longrightarrow SiF_y \tag{4.9}$$

Many important aspects of this mechanism are not at all well understood, such as the extent of fragmentation in (4.4), whether the dissociative chemisorption (4.5-4.6) is a concerted process, the identity of the products desorbed in (4.9), and the fate of the residual carbon from (4.6).

Since the plasma contains ions, electrons, and ionizing radiation, in addition to the reactive neutral species enumerated in the foregoing mechanism, characterization of the details of the surface chemistry becomes very difficult. IRMPD provides a method for cleanly and selectively generating free radicals, particularly from fluorohalocarbon precursors, in the absence of these interfering species. The first report of surface etching by laser-generated free radicals was by Steinfeld et al. [1980]. In that experiment, CF_3Br was dissociated by the 9R(28) line of CO_2 laser, and the products interacted with an adjacent silicon oxide surface. Scanning stylus-probe measurements of surface profiles suggested that approximately 1.3 $\times 10^{13}$ surface atoms were removed by each CO_2 laser pulse. Detailed confirmation of this result had to await experiments in which the surface could be characterized by photoelectron spectroscopy or other quantitative techniques immediately following exposure to the radicals. These measurements are now starting to appear. Roop et al. [1985] have shown, using XPS, that CF_3 radicals undergo partial dissociation on a clean Si surface. The amount of fluorine chemisorbed on the surface follows the IRMPD yield curve for the parent fluorocarbon (C_2F_6, in this case), as is shown in Fig. 4.1. Selamoglu et al. (1986) have used very-low-pressure-photolysis (VLPΦ) to investigate the reaction of CF_3 radicals generated by IRMPD of CF_3I with a silicon oxide surface. In this technique, mass spectrometry is used to measure the products of the surface reaction. It was found that the sticking coefficient for CF_3 on a silicon oxide surface is in the range of 0.0014 - 0.017 between temperatures of 320 - 530 K. Chuang (1981a) has also reported that IRMPE of SF_6 causes that gas to etch silicon; under those experimental conditions, approximately 4×10^{14} Si atoms are removed by each CO_2 laser pulse. XeF_2, which etches silicon by a spontaneous surface chemical reaction [Winters and Coburn, 1979], also has its reactivity increased by CO_2 laser excitation [Chuang, 1981b].

Deposition is the process whereby a thin film of insulating or conducting material is produced on a surface by reaction of a gas-phase species, containing the material to be deposited, at the surface. Deposition, as well as etching, can be induced by IRMPD. An example is deposition of amorphous silicon (a:Si-H). There have been several reports of "CO_2-laser-assisted" deposition of a:Si-H; these typically use a c.w. CO_2 laser to heat SiH_4 or $SiCl_4$ vapor in contact with a substrate [Baranauskas et al., 1980; Bilenchi et al., 1985]. Deposition has been found to occur following IRMPD of substituted silanes $RSiH_3$ (R = SiH_3, C_2H_5, C_4H_9, or C_6H_5) [Francisco et al., 1984]. The reaction which takes place in the case of the alkylsilanes is a four-center elimination

$$R'CH_2CH_2SiH_3 \longrightarrow [\ R'CH\overset{CH_2}{\underset{H}{\cdots\cdots}}SiH_3\ ]^{\ddagger}$$

$$\longrightarrow R'CH=CH_2 + SiH_4 \ . \tag{4.10}$$

The silane thus produced is very "hot" vibrationally, and so can undergo secondary IRMPD to yield silylene (:SiH_2) and hydrogen by the three-center

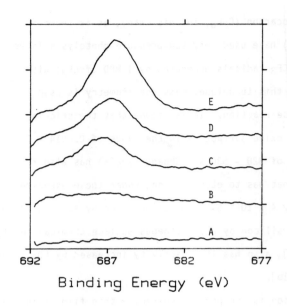

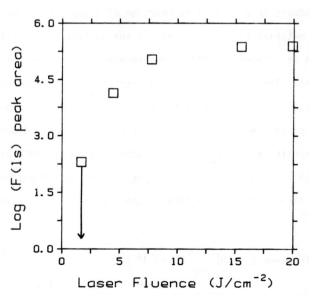

Figure 4.1. Top: F(1s) XPS signal on Si(111)7x7 exposed to C₂F₆ dissociated at varying IR laser fluences, viz., 1.7 J/cm² (A) to 20 J/cm² (E). Bottom: coverage vs. fluence dependence derived from integrated XPS peaks [Roop et al,. 1985].

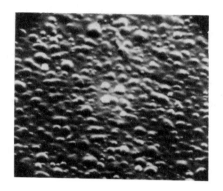

Figure 4.2. Scanning electron microphotographs of amorphous silicon deposits produced by IRMPD of various silanes. Top: phenylsilane on Ag (x2000); Middle: disilane on Ag(x2000); Bottom: disilane on Ag (10,000x) [Francisco et al., 1984].

elimination mechanism noted above [Eq. (4.3)]. The formation of SiH_2 following IRMPD of alkylsilanes has been recently confirmed by laser-induced fluorescence detection of that species [Thoman and Steinfeld, 1985; Rayner et al., 1985].

The SiH_2 thus produced can deposit on adjacent surfaces to form a:Si-H. The morphology of these deposits tends to be quite irregular, with Si-containing nodules forming as shown in Figure 4.2. This may be accounted for by prenucleation of silicon clusters in the gas phase prior to their deposition on the surface.

Most studies of photodeposition of metallic films have employed u.v. (excimer) or visible (Ar^+, Kr^+) laser radiation to decompose the volatile metal-containing species. A recent example of the use of infrared lasers is in deposition of Ni films by focussing the output of a pulsed CO_2 laser into $Ni(CO)_4$ above a substrate [Jervis, 1985]. In this case, however, there is no resonant absorption in the gas at 10.6 μm, and the decomposition process results from laser-induced dielectric breakdown rather than from IRMPD.

5. CONCLUDING REMARKS

We have now nearly completed our survey of infrared multiple-photon phenomena, and the reader, as well as the authors, must be relieved that the end is in sight. We have reviewed some of the historical background, theoretical models and experimental probes for IRMPE, and the most important chemical applications. What conclusions can be drawn from all of this?

One evident conclusion is that measurement of highly averaged (bulk) observables can at best provide only qualitative insight into the IRMPE process. While such insights may be quite valuable for predicting the overall behavior of a system, information on the detailed dynamics is lost. As a result of rapid IVR, an IRMP-activated molecule at any specified energy can be described only in terms of a superposition of eigenstates. Furthermore, the distribution over energies (for specified IR fluence and/or intensity) is not a delta-function, but possesses an appreciable width. In any bulk sample, there is an additional stage of averaging, over the local fluence variations in the volume of the sample. As a result, bulk measurements can be interpreted only in the context of a theoretical model, and most models predict that such measurements will be very insensitive to details of the dynamics.

For this reason, IRMPE is not the method of choice for preparing microscopic quantum states for further dynamical studies, unless the internal distributions can be characterized in considerable detail. We have seen in Section 3.B, however, that this goal has proven to be an elusive one. Indeed, it may not be possible to understand IRMPE dynamics in complete detail, because the spectroscopy at high excitation levels eventually becomes too complex to permit an exact description, and the multistate dynamics can be treated only approximately. To be sure, alternative routes to highly excited vibrational levels have their own limitations. Chemical excitation techniques produce just as poorly defined initial distributions as IRMPE, and the final states produced by stimulated-emission pumping may be very nonrepresentative of the full ground-state potential energy surface.

In light of these conclusions, what have IRMPA, IRMPE, and IRMPD contributed to our understanding of molecular phenomena? Perhaps the most important contribution is that the very existence of these phenomena has forced us to re-examine some of the fundamental premises of molecular dynamics. In particular, the strong evidence for IVR has led to a deeper understanding of "chaotic" (i.e., non-regular) behavior of molecular vibrational motions at sufficiently high energy, and this has in turn been part of a much larger intellectual effort on dissipative phenomena in general.

At a more practical level, IRMPE has found extensive applications in chemical kinetics, particularly for measurement of rate coefficients for free radicals and other transient reactive species.

Under ordinary conditions, there is no mode specificity: all laser-induced chemical reactions appear to involve "known" chemistry. For this reason, no useful direct synthetic applications of IRMPD have appeared. Some limited applications of IRMPD for isotope-selective chemistry may be possible.

IRMPE appears to be a useful technique for initiating chemical reactions in thin films and at surfaces, both for mechanistic studies and for actual surface modification. Although UV (excimer laser) initiation is at least equally important for such applications, there are several systems, such as silane or fluorohalocarbon decomposition, in which IR excitation may be advantageous.

To conclude this discussion, we consider some of the still unanswered questions concerning IR multiple-photon phenomena, and possible approaches to answering them.

First, the nature of the superposition of molecular eigenstates at a specified energy is not well understood. If it were possible to observe the transition from discrete states to the quasicontinuum, our understanding of this process might be improved. A possible approach is to use picosecond or sub-picosecond time resolution to observe these states. Simply exciting at these time scales and monitoring the subsequent reaction will not be sufficient, since the much longer time scale at which the reaction occurs will destroy information about the initially prepared states; it is necessary to employ picosecond sources for both pumping and probing the molecules, possibly using CARS or some other Raman process for the latter.

The role of collisions in IRMPE is not well understood, especially in the QC. Experiments have been carried out which measure average vibrational deactivation from high levels, but little information is available on processes such as dephasing, intramolecular V-V transfer, or collision-induced vibration-rotation coupling.

Finally, it would be very desirable to get away from the 10-μm excitation regime which is imposed by the CO_2 laser. Excitation in the 3 to 4-μm region, corresponding to hydrogen stretching motions, or in overtone bands, could reveal a very different kind of dynamics, as compared with excitation of lower-frequency skeletal vibrations. Progress in this area is limited essentially by the availability of suitable sources; in this context, the availability of picosecond, high-power, tunable IR radiation from the next generation of free-electron lasers may provide the basis for a resurgence of activity in infrared photochemistry and multiple-photon processes.

ACKNOWLEDGMENTS

The work in my laboratory reported in this review has been supported by the National Science Foundation and the Air Force Office of Scientific Research. Dr. Francisco wishes to thank the Office of the Provost at M.I.T. for award of a postdoctoral fellowship. We also thank Profs. N. Bloembergen and M.F. Goodman for permission to include previously published material, Messrs. Sidney Stubbs and Ed Bethel for their hospitality during preparation of this review, and Ms. Karen Moye for her patient, cheerful, and expert word-processing.

REFERENCES

Ackerhalt, J.R., and Galbraith, H.W., 1978, J. Chem. Phys. 69:1200.

Ackerhalt, J.R., and Shore, B.W., 1977, Phys. Rev. A16:277.

Adler-Golden, S.M., and Steinfeld, J.I., 1980, Chem. Phys. Letts. 76:479.

Adler-Golden, S.M., Schweitzer, E.L., and Steinfeld, J.I., 1982, J. Chem. Phys. 76:2201.

Alimpiev, S.S., Valyanskii, S.I., Nikiforov, S.M., Smirnov, V.V., Sartakov, B.G., Fabelinskii, V.I., and Shtarkov, A.L., 1982, JETP Letts. 35:361.

Ambartzumian, R.V. and Letokhov, V.S., 1977a, in Chemical and Biochemical Application of Lasers, Vol. 3 (C.B. Moore, ed.), Academic Press, New York: p. 167.

Ambartzumian, R.V. and Letokhov, V.S., 1977b, Accts. Chem. Res. 10:61.

Ambartzumian, R.V., Gorokhov, Yu.A., Letokhov, V.S., and Makarov, G.N., 1975, JETP Letts. 21:375.

Ambartzumian, R., Furzikov, N., Gorokhov, Yu., Letokhov, V., Makarov, G., and Puretzky, A., 1976a, Opt. Commun. 18:517.

Ambartzumian, R.V., Gorokhov, Y.A., Letokhov, V.S., Makarov, G.N., and Puretzky, A.A., 1976b, JETP Lett., 23:22.

Ambartzumian, R.V., Gorokhov, Yu.A., Letokhov, V.S., and Makarov, G.N., 1976c, Sov. Phys. JETP 42:993.

Ambartzumian, R., Makarov, G., and Puretzky, A., 1977, in Laser Spectroscopy III (J. Hall and J. Carlsten, eds.), Springer-Verlag, Berlin.

Ashfold, M.N.R., Hancock, G., and Ketley, G.W., 1979, Farad. Disc. Chem. Soc. 67:204.

Ashfold, M.N.R., Atkins, C.G., Hancock, G., and MacRobert, A.J., 1984, Israel J. Chem. 24:197.

Avouris, Ph., Loy, M.M.T., and Chan, I.Y., 1979, Chem. Phys. Letts. 63:624.

Bagratashvili, V.N., Knyazev, I.N., Letokhov, V.S., and Lobko, V.V., 1976, Opt. Commun. 18:525.

Bagratashvili, V.N., Dolzhikov, V.S., and Letkhov, V.S., 1979, Sov. Phys. JETP 49:8.

Bagratashvili, V.N., Vainer, Yu.G., Dolzhikov, V.S., Kolyakov, S.F., Makarov, A.A., Malyavkin, L.P., Ryabov, E.A., Sil'kis, E.G., and Titov, V.D., 1980a, JETP Letts. 30:471

Bagratashvili, V.N., Vainer, Yu.G., Doljikov, V.S., Koliakov, S.F., Makarov, A.A., Malyavkin, L.P., Ryabov, E.A., Silkis, E.G., and Titov, V.D., 1980b, Appl. Phys. 22:101.

Bagratashvili, V.N., Vainer, Yu.G., Doljikov, V.S., Letokhov, V.S., Makarov, A.A., Malyavkin, L.P., Ryabov, E.A., and Silkis, E.G., 1981, Optics Letts. 6:148.

Bagratashvili, V.N., Kuzmin, M.V., and Letokhov, V.S., 1984, J. Phys. Chem. 88:5780.

Baklanov, A.V., Molin, Y.N., and Petrov, A.K., 1979, Chem. Phys. Lett. 68:329.

Baranauskas, V., Mammana, C.I.Z., Klinger, R.E., and Greene, J.E., 1980, Appl. Phys. Letts. 36:930.

Barker, J.R., 1980, J. Chem. Phys. 72:3686.

Bauer, S.H., 1978, Chem. Rev. 78:147.

Bäuerle, D. (ed.), 1984, Laser Processing and Diagnostics, Ser. Chem. Phys. No. 39 (Springer-Verlag, Berlin).

Bewick, C.P., Duval, A.B., and Orr, B.J., 1985, J. Chem. Phys. 82:3470.

Bilenchi, R., Gianinoni, I., Musci, M., Murri, R., and Taschetti, S., 1985, Appl. Phys. Letts. (to be published).

Bittenson, S., and Houston, P.L., 1977, J. Chem. Phys. 67:4819.

Black, J.G., Kolodner, P., Shultz, M.J., Yablonovitch, E., and Bloembergen, N., 1979, Phys. Rev. A19:704.

Bloch, F., 1957, Phys. Rev. 105:1206.

Bloembergen, N., 1975, Opt. Commun. 15:416.

Bloembergen, N., Burak, I., Mazur, E., and Simpson, T.B., 1984, Israel J. Chem. 24:179.

Borisevich, N.A., Blinov, S.I., Dorokhin, A.V., Zalesskaja, G.A., Kotov, A.A., 1979, Izv. Acad. Nauk, Ser. Fiz. 43:293.

Brenner, D.M., 1978, Chem. Phys. Lett. 57:357.

Brenner, D.M., 1981, J. Chem. Phys. 74:2293.

Brenner, D.M., Brezinsky, K. and Curtis, D.M., 1980, Chem. Phys. Letts. 72:202.

Brenner, D.M., Spencer, M., and Steinfeld, J.I., 1981, J. Chem. Phys. 75:3153.

Brenner, D.M., Spencer, M.N., and Steinfeld, J.I., 1983, J. Chem. Phys. 78:136.

Brunner, F., Cotter, T.P., Kompa, K.L., and Proch, D., 1977, J. Chem. Phys. 67:1547.

Burak, I., Nowak, A.V., Steinfeld, J.I., and Sutton, D.G., 1969, J. Chem. Phys. 51:2275.

Burak, I., Houston, P., Sutton, D.G., and Steinfeld, J.I., 1970, J. Chem. Phys. 53:3632.

Burak, I., Tsao, J.Y., Prior, Y., and Yablonovitch, E., 1979a, Chem. Phys. Lett. 68:31.

Burak, I., Quelly, T.J., and Steinfeld, J.I., 1979b, J. Chem. Phys. 70:334.

Campbell, J.D., Hancock, G., Halpern, J.B., and Welge, K.H., 1976, Opt. Commun. 17:38.

Cantrell, C.D., 1977, in Laser Spectroscopy III, (J. Hall and J.L. Carlsten, eds.) Springer, New York: p. 109.

Cantrell, C.D., and Galbraith, H.W., 1977, Opt. Commun. 21:374.

Cantrell, C.D., Galbraith, H.W., and Ackerhalt, J.R., 1978, in Multiphoton Processes (J.H. Eberly and P. Lambropulous, eds.) Wiley, New York: p. 331.

Cantrell, C.D., Freund, S.M., and Lyman, J.L., 1979, in Laser Handbook Vol. 3 (M. Stitch, ed.) North-Holland Publishing Co., Amsterdam: pp. 485-576.

Carmeli, B., and Nitzan, A., 1980a, J. Chem. Phys. 72:2054.

Carmeli, B., and Nitzan, A., 1980b, J. Chem. Phys. 72:2070.

Chekalin, N.V., Dolzhikov, V.S., Letokhov, V.S., and Shibonov, A.N., 1977, Appl. Phys. 12:191.

Chuang, T.J., 1981a,b, J. Chem. Phys. 74:1453, 1461.

Chuang, T.J., 1982, J. Vac. Sci. Tech. 21:798.

Corkum, R., Willis, C., and Back, R.A., 1977, Chem. Phys. 24:13.

Dai, H.-L., Kung, A.H., and Moore, C.B., 1979, Phys. Rev. Letts. 43:761.

Dai, H.-L., Kung, A.H., and Moore, C.B., 1980, J. Chem. Phys. 73:6124.

Dai, H.-L., Specht, E., Berman, M.R., and Moore, C.B., 1982, J. Chem. Phys. 77:4494.

Dai, H.-L., Field, R.W., and Kinsey, J.L., 1985a, J. Chem. Phys. 82:1606.

Dai, H.-L., Korpa, C.L., Kinsey, J.L., and Field, R.W., 1985b, J. Chem. Phys. 82:1688.

Danen, W.C., and Jang, J.C., 1981, in Laser-Induced Chemical Processes (J.I. Steinfeld, ed.), Plenum Publishing Co., New York: pp. 45-164.

164

Danen, W.C., Rio, V.C., and Setser, D.W., 1982, J. Amer. Chem. Soc. 104:5431.

Deutsch, T., 1977, Opt. Lett. 1:25.

Dever, D.F., and Grunwald, E., 1976, J. Amer. Chem. Soc. 98:5055.

Dilonardo, M., 1981, Il Nuovo Cimento, 63B:131.

Dove, J.E., Hippler, H., and Troe, J., 1985, J. Chem. Phys. 82:1907.

Dübal, H.-R., and Crim, F.F., 1985, J. Chem. Phys. 83:3863.

Dubs, M., Harradine, D., Schweitzer, E., Steinfeld, J.I., and Patterson, C., 1982, J. Chem. Phys. 77:3824.

Duperrex, R., and van der Bergh, H., 1979, J. Chem. Phys. 71:3613.

Eberhardt, J.E., Knott, R.B., Pryor, A.W., and Gilbert, R.G., 1982, Chem. Phys. 69:45.

Eleckij, A.V., Klimov, V.D., and Legasov, V.A., 1977, Dokl. Akad. Nauk SSSR 237:1396.

Esherick, P., Grimley, A.J., and Owyoung, A., 1982, Chem. Phys. 73:271.

Evans, D.K., McAlpine, R.D., and Adams, H.M., 1982, J. Chem. Phys. 77:3551.

Evans, D.K., McAlpine, R.D., and Adams, H.M., 1984, Israel J. Chem. 24:187.

Fang, H.L., and Swofford, R.L., 1983, in Ultrasensitive Laser Spectroscopy (D.S. Kliger, ed.) Academic Press, New York: pp. 175-232.

Farneth, W.E., Thomsen, M.W., and Berge, M.A., 1979, J. Amer. Chem. Soc. 101:6468.

Feldmann, D., Zacharias, H., and Welge, K.H., 1980, Chem. Phys. Letts. 69:466.

Feldmann, D., Laukemper, J., and Welge, K.H., 1983, J. Chem. Phys. 79:278.

Felker, P.M., and Zewail, A.H., 1985a,b,c, J. Chem. Phys. 82:2961, 2975, 2994.

Felker, P.M., Lambert, W.R., and Zewail, A.H., J. Chem. Phys. 82:3003.

Filseth, S.V., Hancock, G., Fournier, J., and Meier, K., 1979, Chem. Phys. Lett. 61:288.

Francisco, J.S., and Steinfeld, J.I., 1981, Int. J. Chem. Kinet. 13:615.

Francisco, J.S., Steinfeld, J.I., and Gilbert, R.G., 1981a, Chem. Phys. Lett. 82:311.

Francisco, J.S., Finders, M.A., and Steinfeld, J.I., 1981b, Int. J. Chem. Kinet. 13:627.

Francisco, J.S., Lawrance, W.D., Steinfeld, J.I., and Gilbert, R.G., 1982, J. Phys. Chem. 86:724.

Francisco, J.S., Zhu Qingshi, and Steinfeld, J.I., 1983, J. Chem. Phys. 78:5339.

Francisco, J.S., Joyce, S.A., Steinfeld, J.I., and Walsh, F., 1984, J. Phys. Chem. 88:3098.

Frankel, D.S., Jr., 1976, J. Chem. Phys. 65:1696.

Frankel, D., and Manuccia, T., 1978, Chem. Phys. Letts. 54:451.

Frei, H., and Pimentel, G., 1983, J. Chem. Phys. 78:3698.

Frei, H., and Pimentel, G., 1985, Ann. Rev. Phys. Chem. 36:491.

Friedmann, H., and Ahiman, V., 1980, Opt. Commun. 33:163.

Fuss, W., 1979a, Chem. Phys. 36:135.

Fuss, W., 1979b, V. Chem. Phys. 68:135.

Fuss, W., 1980, Chem. Phys. Letts. 71:77.

Fuss, W., 1982, J. Phys. Chem. 86:731.

Fuss, W., and Cotter, T.P., 1977, Appl. Phys. 12:265.

Fuss, W., Hartmann, J., and Schmidt, W.E., 1978, Appl. Phys. 15:297.

Galbraith, H.W., and Ackerhalt, J.R., 1981, in Laser-Induced Chemical Processes (J.I. Steinfeld, ed.) Plenum, New York: p. 1.

Gillespie, D.T., 1976, J. Comput. Phys. 22:403.

Gillespie, D.T., 1977, J. Phys. Chem. 81:2340.

Glatt, I., and Yogev, A., 1981, Chem. Phys. Letts. 77:228.

Gough, T.E., Miller, R.E., and Scoles, G., 1978, in Advances in Laser Chemistry (A.H. Zewail, ed.) Springer-Verlag, Berlin: p. 433.

Gower, M.C., and Billman, K.W., 1977a, Appl. Phys. Lett., 30:514.

Gower, M.C., and Billman, K.W., 1977b, Chem. Phys. Lett., 20:123.

Gower, M.C., and Gustafson, T., 1977, Opt. Commun. 23:69.

Gozel, P., Calpini, B., and van den Bergh, H., 1984, Israel J. Chem. 24:210.

Grant, E.R., Coggiola, M.J., Lee, Y.T., Schulz, P.A., Sudbo, Aa.S., and Shen, Y.R., 1977, Chem. Phys. Lett. 52:595.

Grant, E.R., Schulz, P.A., Sudbo, Aa.S., Shen, Y.R., and Lee, Y.T., 1978, Phys. Rev. Lett. 40:115.

Grimley, A.J., and Houston, P.L., 1980, J. Chem. Phys. 72:1471.

Grimley, A.J., and Stephenson, J.C., 1981, J. Chem. Phys. 74:447.

Grunwald, E, Dever, D.F., and Keehn, P.M., 1978, Megawatt Infrared Laser Chemistry (Wiley, New York).

Hackett, P.A., Gauthier, M., Willis, C., and Pilon, R., 1979, J. Chem. Phys. 71:546.

Hackett, P.A., Willis, C., Gauthier, M., and Alcock, A.J., 1984, in Applications of Lasers to Industrial Chemistry (R.L. Woodin and A. Kaldor, eds.) SPIE 458:65.

Ham, D.O., and Rothschild, M., 1977, Opt. Lett. 1:28.

Hamilton, C., Field, R.W., and Kinsey, J.L., 1986, Ann. Rev. Phys. Chem. 37: in press.

Hancock, G., Hennesy, R.J., and Villis, T., 1979, J. Photochem. 10:305.

Harradine, D., Foy, B., Laux, L., Dubs, M., and Steinfeld, J.I., 1984, J. Chem. Phys. 81:4267.

Hartford, A., 1978, Chem. Phys. Lett., 57:352.

Haub, J.G., 1985, Ph.D. Thesis, University of New South Wales (Sydney, Australia).

Haub, J.G., and Orr, B.J., 1984, Chem. Phys. Letts. 107:162.

Heilweil, E.J., Casassa, M.P., Cavanagh, R.R., and Stephenson, J.C., 1985, J. Chem. Phys. 82:5216.

Herman, I.P., Magnotta, F., and Aldridge, F.T., 1984, Israel J. Chem. 24:192.

Heymann, M., Hippler, H., and Troe, J., 1984, J. Chem. Phys. 80:1853.

Hippler, H., Troe, J., and Wendelken, H.J., 1983a, J. Chem. Phys. 78:6709.

Hippler, H., Troe, J., and Wendelken, H.J., 1983b, J. Chem. Phys. 78:6718.

Hippler, H., Lindemann, L., and Troe, J., 1985, J. Chem. Phys. 83:3906.

Hodgkinson, D.P., and Briggs, J.S., 1976, Chem. Phys. Lett. 43:451.

Hodgkinson, D.P., and Briggs, J.S., 1977, J. Phys. B10:2583.

Horsley, J., Stone, J., Goodman, M., and Dows, D., 1979, Chem. Phys. Letts. 43:451.

Horwitz, A., Preses, J.M., Weston, R.E., Jr., and Flynn, G., 1981, J. Chem. Phys. 74:5008.

Houston, P.L., 1981, Adv. Chem. Phys. 47:625.

Houston, P.L, Nowak, A.V., and Steinfeld, J.I., 1973, J. Chem. Phys. **58:3373.**

Hudgens, J.W., 1978, J. Chem. Phys. 68:777.

Hudgens, J.W., and McDonald, J.D., 1985, in Lasers as Reactants and Probes in Chemistry (W.M. Jackson and A.B. Harvey, eds.), Howard University Press, Washington, D.C.: p. 71.

Hwang, W.C., and Sandstrom, J.K., 1981, "Synthesis of hydrazine by IR laser photodissociation", Report ATR-81(8420)-1, Aerospace Corp., El Segundo, Calif.

Isenor, N., and Richardson, M.C., 1971, Appl. Phys. Letts. 18:224.

Isenor, N., Merchant, V., Hallsworth, R., and Richardson, M., 1973, Can. J. Phys. 51:1281.

Jalenak, W.A., and Nogar, N.S., 1980, J. Phys. Chem. 84:2993.

Jasinski, J.M., and Estes, R.D., 1985, Chem. Phys. Letts. 117:495.

Jensen, C.C., Anderson, T.G., Reiser, C., and Steinfeld, J.I., 1979, J. Chem. Phys. 71:3648.

Jervis, T.R., 1985, J. Appl. Phys. 58:1400.

Keefer, D.R., Allen, J.E. Jr., and Person, W.B., 1976, Chem. Phys. Lett. 43:394.

Kim, K.C., Freund, S.M., Sorem, M.S., and Smith, D.F., 1985, J. Chem. Phys. 83:4344.

Kimel, S., and Speiser, S., 1977, Chem. Rev. 77:437.

King, D.S., and Stephenson, J.C., 1977, Chem. Phys. Letts. 47:36.

King, D.S., and Stephenson, J.C., 1985, J. Chem. Phys. 82:2236.

Knudsen, A.K., and Pimentel, G., 1983, J. Chem. Phys. 78:6780.

Kolodner, P., Winterfeld, C., and Yablonovitch, E., 1977, Opt. Commun. 20:119.

Kolodner, P., Kwok, H.S., Black, J.G., and Yablonovitch, E., 1979, Opt. Lett. 4:38.

Kompa, K.L., Fuss, W., Proch, D., and Schmid, W.E., 1978, in Electronic and Atomic Collisions, (E. Watel, ed.) North Holland Publ., Amsterdam: p. 737.

Kudriavtsev, Yu.A., and Letokhov, V.S., 1980, Chem. Phys. Letts. 50:353.

Kulp, T., Ruoff, R., Stewart, G., and McDonald, J.D., 1984, J. Chem. Phys. 80:5359.

Kulp, T., Ruoff, R.S., and McDonald, J.D., 1985, J. Chem. Phys. 82:2175.

Kwok, H.S., and Yablonovitch, E., 1978, Phys. Rev. Lett. 41:745.

Kwok, H.S., Yablonovitch, E., and Bloembergen, N., 1981, Phys. Rev. A23:3094.

Kyrölä, E., and Eberly, J.H., 1985, J. Chem. Phys. 82:1841.

Lamb, W., 1977, in Laser Spectroscopy III (J. Hall and J. Carlsten, eds.), Springer-Verlag, Berlin: p. 116.

Langmuir, I., 1920, J. Am. Chem. Soc. 42:2190.

Larsen, D.M., 1976, Opt. Commun. 19:404.

Larsen, D.M., and Bloembergen, N., 1976, Opt. Commun. 17:254.

Lawrance, W.D., Knight, A.E.W., Gilbert, R.G., and King, K.D., 1981a, Chem. Phys. 56:343.

Lawrance, W.D., Silverstein, J., Zhang Fu-Min, Zhu Qingshi, Francisco, J.S., and Steinfeld, J.I., 1981b, J. Phys. Chem. 85:1961.

Laux, L., Foy, B., Harradine, D., and Steinfeld, J.I., 1984, J. Chem. Phys. 80:3499.

Lesiecki, M.L., Hicks, K.W., Orenstein, A., and Guillory, W.A., 1980, Chem. Phys. Letts. 71:72.

Letokhov, V.S., and Moore, C.B., 1977, in Chemical and Biological Applications of Lasers, Vol. 3 (C.B. Moore, ed.), Academic Press, New York: pp. 1-165.

Levy, M.R., Reisler, H., Mangir, M.S., and Wittig, C., 1980, Opt. Eng. 19:29.

Lussier, F.M., and Steinfeld, J.I., 1977, Chem. Phys. Letts. 50:175.

Lussier, F.M., Steinfeld, J.I., and Deutsch, T.F., 1978, Chem. Phys. Letts. 58:277.

Lyman, J.L., Jensen, R.J., Rink, J., Robinson, C.P., and Rockwood, S.D., 1975, Appl. Phys. Letts. 27:87.

Lyman, J.L., Rockwood, S.D., and Freund, S.M., 1977, J. Chem. Phys. 67:4545.

Lyman, J.L., Radziemski, L.J., Jr., and Nilsson, A.C., 1980, IEEE J. Quantum Electronics QE-16:1174.

Lyman, J.L., Quigley, G.P., and Judd, O.P., 1981, in Multiple Photon Excitation and Dissociation of Polyatomic Molecules (C. Cantrell, ed.), Springer-Verlag, Berlin.

Lyman, J.L., Muller, G., Houston, P.L., Piltch, M., Schmid, W.E., and Kompa, K.L., 1985, J. Chem. Phys. 82:810.

Malatesta, V., Willis, C., and Hackett, P.A., 1984, U.S. Patent No. 4,447,304.

Marling, J.B., Herman, I.P., and Thomas, S.J., 1980, J. Chem. Phys. 72:5603.

Martinez, R.I., Huie, R.E., Herron, J.T., and Braun, W., 1980, J. Phys. Chem. 84:2344.

Mazur, E., Burak, I., and Bloembergen, N. 1984, Chem. Phys. Letts. 105:258.

McAlpine, R.D., Evans, D.K., and Adams, H.M., 1983, J. Chem. Phys. 78:5990.

McMillen, D.F., Lewis, K.E., Smith, G.P., and Golden, D.M., 1982, J. Phys. Chem. 86:709.

Medvedev, E.S., 1979, Chem. Phys. 44:103.

Morrison, R.J.S., Loring, R.F., Furlex, R.L., and Grant, E.R., 1981, J. Chem. Phys. 75:148.

Moulton, P.F., Larsen, D.M., Walpole, J.N., and Mooradian, A., 1977, Opt. Lett. 1:51.

Mukamel, S., and Jortner, J., 1976a, J. Chem. Phys. 65:5204.

Mukamel, S., and Jortner, J., 1976b, Chem. Phys. Letts. 40:150.

Mukamel, S., 1979a, Phys. Rev. Lett. 42:168.

Mukamel, S., 1979b, J. Chem. Phys. 70:2419.

Mukamel, S., 1979c, J. Chem. Phys. 71:2012.

Narducci, L., Mitra, S., Shatas, R., and Coulter, C., 1977, Phys. Rev. A. 16:247.

Neudorfl, P., Jodhan, A., and Strauz, O.P., 1980, J. Phys, Chem. 84:338.

Newman, C.G., O'Neal, H.E., Ring, M.A., Leska, F., and Shipley, N., 1979, Intern. J. Chem. Kinet. 11:1167.

Okafa, E.N. and Whittle, E., 1975, Int. J. Chem. Kinet. 7:273.

Oppenheim, I., Shuler, K., and Weiss, G.H., 1977, Stochastic Processes in Chemical Physics: The Master Equation (M.I.T. Press, Cambridge, Mass.).

Orr, B.J., and Nutt, G.F., 1980, J. Mol. Spectroscopy 84:272.

Orr, B.J., and Haub, J.G., 1984, J. Mol. Spectroscopy 103:1.

Orr, B.J., Haub, J.G., Nutt, G.F., Steward, J.L., and Vozzo, O., 1981, Chem. Phys. Letts. 78:621.

Orr, B.J., Haub, J.G., and Haines, R., 1984, Chem. Phys. Letts. 107:168.

Osgood, R.M., Jr., 1983, Ann. Rev. Phys. Chem. 34:77.

Padrick, T.D., Hays, K., and Palmer, M.A., 1980, Chem. Phys. Letts. 70:63.

Patterson, C.W., McDowell, R.S., Moulton, P.F., and Mooradian, A., 1981, Opt. Letts. 6:93.

Perrin, F., 1919, Ann. Phys. 11:5.

Porter, G., and Steinfeld, J.I., 1966, J. Chem. Phys. 45:3456.

Preses, J.M., and Flynn, G.W., 1977, J. Chem. Phys. 66:3112.

Preses, J.M., Weston, R.E., Jr., and Flynn, G.W., 1977, Chem. Phys. Lett. 46:60.

Presser, N., Mao, C.-R., Moriarty, R.M., and Gordon, R.J., 1983a, J. Chem. Phys. 78:6021.

Presser, N., Barker, J.R., and Gordon, R.J., 1983b, J. Chem. Phys. 78:2163.

Proch, D., and Schröder, H., 1979, Chem. Phys. Letts. 61:426.

Purnell, J.H., and Walsh, R., 1966, Proc. Roy. Soc. A293:543.

Quack, M., 1978, J. Chem. Phys. 69:1282.

Quack, M., 1982, in Adv. Chem. Phys. Vol. 50: Dynamics of the Excited State (K. Lawley, ed.), J. Wiley Ltd., Chichester: pp. 395-473. Quick, C.R., Jr., and Wittig, C., 1978, J. Chem. Phys. 69:4201.

Quick, C.R., Jr., and Wittig, C., 1978, J. Chem. Phys. 69:4201.

Rayner, D.M., Steer, R.P., Hackett, P.A., Wilson, C., and John, P., 1983, Chem. Phys. Letts. (to be published).

Reddy, K.V., 1984, in Applications of Lasers to Industrial Chemistry (R.L. Woodin and A. Kaldor, eds.) SPIE 458:53.

Reddy, K.V., Heller, D.F., and Berry, M.J., 1982, J. Chem. Phys. 76:2814.

Reiser, C., and Steinfeld, J.I., 1980, Opt. Eng. 19:2.

Reiser, C., Lussier, F.M., Jensen, C.C., and Steinfeld, J.I., 1979, J. Amer. Chem. Soc. 101:350.

Reiser, C., Steinfeld, J.I., and Galbraith, H.W., 1981, J. Chem. Phys. 74:2189.

Richardson, T.H., and Setser, D.W., 1977, J. Phys. Chem. 81:2301.

Ringwelski, A.Z., Presser, N., and Gordon, R.J., 1984, Israel J. Chem. 24:214.

Robertson, R., Hils, D., and Gallagher, A., 1984, Chem. Phys. Letts. 103:397.

Robinson, P.J., and Holbrook, K.A., 1972, <u>Unimolecular Reactions,</u> Wiley-Interscience, New York.

Roop, B., Joyce, S., Schultz, J.C., and Steinfeld, J.I., 1985, J. Chem. Phys. 83:6012.

Rosenfeld, R.N., Brauman, J.I., Baker, J.R., and Golden, D.M., 1977, J. Amer. Chem. Soc. 99:8063.

Rossi, M.J., Barker, J.R., and Golden, D.M., 1979, J. Chem. Phys. 71:3722.

Rossi, M.J., Barker, J.R., and Golden, D.M., 1982, J. Chem. Phys. 76:406.

Rossi, M.J., Pladziewicz, J.R., and Barker, J.R., 1983, J. Chem. Phys. 78:6695.

Schek, I., and Jortner, J., 1979, J. Chem. Phys. 70:3016.

Schulz, P.A., Sudbϕ, Aa.S., Krajnovich, D.J., Kwok, H.S., Shen, Y.R., and Lee, Y.T., 1979, Ann. Rev. Phys. Chem. 30:379.

Schulz, P.A., Sudbϕ, A.S., Grant, E.R., Shen, Y.R., and Lee, Y.T., 1980, J. Chem. Phys. 72:4985.

Schweitzer, E., 1985, Ph.D. Thesis, Massachusetts Institute of Technology.

SCM Specialty Chemicals Co., 1984, Research Chemicals Catalog: p. 8.

Selamoglu, N., Rossi, M.J., and Golden, D.M., 1986, J. Phys. Chem. (to be published).

Selwyn, G.S., and Lin, M.C., 1985, in <u>Lasers as Reactants and Probes in Chemistry</u> (W.M. Jackson and A.B. Harvey, eds.), Howard University Press, Washington, D.C.: p. 269.

Shaub, W.M., and Bauer, S.H., 1975, Intern. J. Chem. Kinet. 7:509.

Shultz, M.J., and Yablonovitch, E., 1978, J. Chem. Phys. 68:3007.

Sibert III, E.L., Reinhardt, W.P., and Hynes, J.T., 1984a, J. Chem. Phys. 81:1115.

Sibert III, E.L., Hynes, J.T., and Reinhardt, W.P., 1984b, J. Chem. Phys. 81:1135.

Simpson, T.B., Black, J.G., Burak, I., Yablonovitch, E., and Bloembergen, N., 1985, J. Chem. Phys. 83:628.

Sinha, A., and Kinsey, J.L., 1983, J. Chem. Phys. 80:2029.

Slagle, I.R., and Gutman, D., 1982, J. Amer. Chem. Soc. 104:4741.

Speiser, S., and Jortner, J., 1976, Chem. Phys. Lett. 44:399.

Stafast, H., Opitz, J., and Huber, J.R., 1981, Chem. Phys. 56:63.

Steinfeld, J.I., 1981, in Laser-Induced Chemical Processes (J.I. Steinfeld, ed.), Plenum Publishing Co., New York: pp. 243-267.

Steinfeld, J.I., 1983a, Proceedings, U.S.-Australia Workshop on Reactivity and Energy Distribution in Surface-Active Free Radicals (R.G. Gilbert, ed.): pp. 5-12.

Steinfeld, J.I., 1983b, in Energy Storage and Redistribution in Molecules (J. Hinze, ed.), Plenum Press, New York: pp. 1-15.

Steinfeld, J.I., Burak, I., Sutton, D.G., and Nowak, A.V., 1970, J. Chem. Phys. 52:5421.

Steinfeld, J.I., Anderson, T.G., Reiser, C., Denison, D.R., Hartsough, L.D., and Hollahan, J.R., 1980, J. Electrochem. Soc. 127:514.

Stephenson, J.C., and King, D.S., 1978, J. Chem. Phys. 69:1485.

Stephenson, J.C., and King, D.S., 1983, J. Chem. Phys. 78:1867.

Stephenson, J.C., King, D.S., Goodman, M.F., and Stone, J., 1979, J. Chem. Phys. 70:4496.

Stephenson, J.C., Blazy, J.A., Li, C-L., and King, D.S., 1982, J. Chem. Phys. 76:5989.

Stewart, G., Ruoff, R., Kulp, T., and Mcdonald, J.D., 1984, J. Chem. Phys. 80:5353.

Stone, J., and Goodman, M.F., 1979, J. Chem. Phys. 71:408.

Stone, J., and Goodman, M.F., 1985, J. Phys. Chem. 89:1250.

Thiele, E., Goodman, M.F., and Stone, J., 1980, Opt. Eng. 19:10.

Thoman, J.W., Jr., and Francisco, J.S., 1985, unpublished results.

Thoman, J.W., Jr., and Steinfeld, J.I., 1985, Chem. Phys. Letts. (to be published).

Troe, J., 1980, J. Chem. Phys. 73:3205.

Tsao, J., Black, J.G., Yablonovitch, E., and Burak, I., 1980, J. Chem. Phys. 73:2076.

Tsao, J.Y., Simpson, T.B., Bloembergen, N., and Burak, I., 1982, J. Chem. Phys. 77:1274.

Walker, R., and Preston, R., 1977, J. Chem. Phys. 67:2017.

West, G.A., Weston, R.E. Jr., and Flynn, G.W., 1978, Chem. Phys. 35:275.

Weston, R.E., Jr., 1982, J. Phys. Chem. 86:4864.

Wilcox, L.R., and Lamb, W.E. Jr., 1960, Phys. Rev. 119:1915.

Winters, H.F., 1978, J. Appl. Phys. 49:5165.

Winters, H.F., and Coburn, J., 1979, Appl. Phys. Letts. 34:70.

Wolfrum, J., 1977, Ber. Bunsenges. Phys. Chem. 81:114.

Wolfrum, J., and Schneider, M., 1984, in Applications of Lasers to Industrial Chemistry (R.L. Woodin and A. Kaldor, eds.), SPIE 458:46.

Wolk, G.L., Weston, R.E., Jr., and Flynn, G.W., 1980, J. Chem. Phys. 73:1649.

Wurzberg, E., and Houston, P.L., 1980, J. Chem. Phys. 72:4811.

Wurzberg, E., Kovalenko, L.J., and Houston, P.L., 1978a, Chem. Phys. 35:317.

Wurzberg, E., Grimley, A.J., and Houston, P.L., 1978b, Chem. Phys. Letts. 57:373.

Yahav, G., and Haas, Y., 1978, Chem. Phys. 35:41.

Yamada, F., Slagle, I.R., and Gutman, D., 1981, Chem. Phys. Letts. 83:409.

Yuan, R.C.L., Preses, J.M., Flynn, G.W., and Ronn, A.M., 1973, J. Chem. Phys. 59:6128.

Zhang Famin, Francisco, J.S., and Steinfeld, J.I., 1982, J. Phys. Chem. 86:2402.

3. DYNAMICS AND SYMMETRIES IN INTENSE MULTIPHOTON PROCESSES: FLOQUET THEORETICAL APPROACHES

Shih-I Chu*

Department of Chemistry
University of Kansas†
Lawrence, Kansas 66045, USA

and

Joint Institute for Laboratory Astrophysics
University of Colorado

and

National Bureau of Standards
Boulder, Colorado 80309, USA

Several recently developed Floquet theoretical approaches for *ab initio* studies of (i) multiphoton excitation of molecules, (ii) SU(N) dynamical evolution and nonlinear coherence of N-level systems in polychromatic fields, and (iii) laser-induced multi-charged ion-atom collisions are reviewed.

*JILA Visiting Fellow (1985).
†Permanent address.

Contents

I. Introduction

The study of nonlinear multiphoton dynamics and the interplay between atomic/molecular structure and intense electromagnetic fields is a
subject of much current interest both theoretically and experimentally.
Although the qualitative features of the multiphoton ionization (MPI),
excitation (MPE), and dissociation (MPD) appear to be established,[1,2]
quantitative analysis of the underlying quantum dynamics is still far
from satisfactory. The major obstacles arise from the complexity of the
molecular level structures and the lack of suitable theoretical and numerical methods for the treatment of molecule-field interaction of large
dimensionality.

The theory of multiphoton processes can be constructed in a fully
quantum mechanical or semiclassical fashion. In the former approach,
both the system and the field are treated quantum mechanically, while in
the semiclassical approach the system is treated quantum mechanically
and the field classically. The semiclassical approach usually leads to
Hamiltonians that are periodic in time through an assumed sinusoidal time
variation of the electromagnetic field. In addition to being considerably
more tractable, the semiclassical theory has been shown to lead to results
that are equivalent to those obtained from the fully quantized theory in
strong fields.[3,4] Indeed much recent multiphoton research has been performed in terms of the semiclassical framework, particularly the Floquet
theoretical approach.[5,6] In this article, we review several recent developments in Floquet theory and quasi-energy methods for _ab initio_ study
of the dynamics and symmetries in intense field multiphoton processes in
gas phase that have been carried out at the University of Kansas. Detailed discussions of other Floquet developments can be found in Ref. 6.

II. Multiphoton Excitation of Small Molecules in Intense
Monochromatic Fields

A. Conventional Floquet Matrix Method

The solutions of linear differential equations with periodic coefficients were first considered by Floquet[7] and Poincaré[8] about a century ago. The Floquet theorem was later used by Autler and Townes[9] to obtain wave functions for the two-level system in terms of infinite continued fractions. Application of Floquet theory to a quantum system began to grow only after the mid 1960's.[3,10,11] In particular, Shirley[3] reformulated the time-dependent problem of the interaction of a two-level quantum system with a strong oscillating classical field as an equivalent time-independent infinite-dimensional Floquet matrix. While this is a semi-classical theory without explicit field quantization, Shirley showed that the Floquet states can be interpreted physically as quantum field states. In fact, his Floquet quasienergy diagram is identical to the dressed-atom picture introduced by Cohen-Tannoudji and Haroche.[12]

Exact analytical solution of the time-dependent Schrödinger equation with temporally periodic Hamiltonian is possible only in exceptional cases. Thus it is in general necessary to develop approximate methods for the treatment of multiphoton excitation (MPE) of atoms and molecules.

Shirley's Floquet matrix method, which is equivalent to summing the entire perturbation expansion to all orders,[13] can be extended to the nonperturbative study of rovibrational multiphoton excitation of a molecule within a single electronic state. In the electric dipole approximation, the interaction potential energy between the quantum system and the classical field is given by

$$V(r,t) = -\vec{\mu}(\vec{r}) \cdot \vec{E}_0 \cos(\omega t + \delta) \quad , \qquad (1)$$

where μ is the dipole moment operator of the molecule, E_0 and δ are the field amplitude and phase, respectively, assumed to be independent of time, and $\vec{r}$ represents the internuclear coordinates. The Floquet matrix possesses a block tridiagonal form as shown in Fig. 1. The determination of the vibrational-rotational quasi-energies and quasi-energy states (QES) thus reduces to the solution of a time-independent Floquet matrix eigen-problem. Figure 1 shows that H_F has a periodic structure with only the number of ω's in the diagonal elements varying from block to block. This structure endows the quasi-energy eigenvalues and eigenvectors of H_F with periodic properties.

Much information can be obtained from the plot of the quasi-energy eigenvalues (or the characteristic exponents) of the Floquet Hamiltonian. The main feature of the quasi-energy plot is illustrated in Fig. 2 for the case of the HF molecule subjected to both the ac and dc fields.[14] Non-linear effects such as power broadening, dynamical Stark effect, Autler-Townes multiplet splitting, hole burning, and S-hump behaviors, etc., are observed and can be correlated with quasi-energy diagrams (Fig. 2). Many of the salient features in the spectral line shapes may be qualitatively understood in terms of an analytical three- or four-level model.[15] The addition of a dc electric field spoils the restriction of the rotational dipole selection rule and induces significant intermixing of the bare molecular vibrator states. Owing to the greater number of strongly coupled nearby states in the dc field, nonlinear optical effects such as those mentioned above appear at a much lower ac field strength than they would in the absence of the dc field. The introduction of an external dc field, therefore, strongly enhances the MPE probabilities and results in a much richer spectrum, in accord with the experimental observations.[16,17]

Shirley's Floquet Hamiltonian method, though powerful, is subject to some practical limitations. In dealing with molecular multiphoton problems, the Floquet matrix is usually truncated to N by N dimensions where $N = N_F N_V N_J$, N_F is the number of Floquet photon blocks, N_V the number of vibrational levels included in one Floquet block, and N_J the number of rotational states included in one vibrational level (Fig. 1). As N increases rapidly with the size of the molecule and with the order of multiphoton processes, and as the computational expense grows as N^3, the full Floquet analysis can become prohibitively costly even for simple diatomic molecules. One is thus led to seek new approximate techniques capable of providing sufficiently accurate results yet involving much smaller Floquet matrix manipulations. The situation here is similar to the field of molecular collision theory: where a variety of angular momentum decoupling methods (for recent reviews, see Ref. 18) have been developed in the past decade to alleviate the problem of a large number of coupled equations. In Secs. IIB and IIC, two of the recently developed approximate Floquet treatments are described.

B. Nonadiabatic Theory for Resonant Multiphoton Absorption

Consider the process of IR multiphoton absorption (MPA) in molecules. The laser frequency is in general of the same order of magnitude as the molecular vibrational frequency and considerably exceeds the rotational frequency. More explicitly, since the Hamiltonian contains fast (vibrational motion and electric dipole interaction) and slow (rotational motion) parts, one anticipates the wave function will have the adiabatic form of the product of a rapidly oscillating function times a function slowly varying in time. Similar to the Born-Oppenheimer approximation, one can determine the rapidly oscillating part by considering the Schrödinger equation with the rotational Hamiltonian frozen. This gives rise to an effective quasi-vibrational energy (QVE) depending parametrically upon the orientation of the molecule. The slowly varying parts of the wave function can then be determined by using the effective QVE as the potential energy for the rotational motion. This adiabatic picture has been used previously[19] to obtain approximate analytical solutions for the rotational spectra of diatomic molecules in some special cases. It provides an adequate zeroth-order description and a convenient basis for the development of an economical and feasible approach for MPA studies. However, the adiabatic approximation breaks down nearby resonant transition regions and nonadiabatic angular couplings among nearly degenerate adiabatic QVE states are required for a proper MPA analysis.[20] The essence of the nonadiabatic Floquet theory is illustrated for the case of a diatomic molecule below. Extension to polyatomic molecules is straightforward.

The Schrödinger equation for a heteronuclear diatomic molecule interacting with a coherent monochromatic field $\vec{E} = \vec{E}_0 \cos(\omega t + \delta)$ has the following form ($\hbar=1$):

$$i\partial\Psi(\vec{r},t)/\partial t = H(\vec{r},t)\Psi(\vec{r},t) \quad , \tag{2}$$

where

$$H(\vec{r},t) = H_v^{(0)}(\vec{r}) + H_{RR}(\hat{r}) - \mu(r)E_0 \cos\theta \cos(\omega t+\delta) \quad . \tag{3}$$

Here $H_v^{(0)}(\vec{r})$ is the diatom vibrational Hamiltonian and $H_{RR}(\hat{r})$ is the rigid rotator Hamiltonian. The vibrational-rotational interaction has been ignored for simplicity. $\mu(r)$ is the molecular electric dipole moment and θ is the angle between the direction of the field $(\vec{E}_0//\vec{z})$ and the axis of the molecule. As the interaction energy is independent of the azimuth angle ϕ, the rotational magnetic quantum number m_j is a constant motion. In the adiabatic approximation, one first solves for the QVE states with the molecular orientation frozen. The resulting Schrödinger equation (with θ fixed) is

$$i\partial\Phi(\vec{r},t,\theta)/\partial t = [H_v^{(0)}(\vec{r})-\mu(r)E_0 \cos\theta \cos(\omega t+\delta)]\Phi(\vec{r},t,\theta) \quad . \tag{4}$$

Corresponding to Eq. (4), an equivalent time-independent Hamiltonian $H_F^{(v)}$ may be written in analogy with the semiclassical Floquet Hamiltonian of Shirley. The resulting matrix block structure [with the dimension only of $(N_F N_v)$ by $(N_F N_v)$] is shown in Fig. 3. The QVE solution $\Phi_{\alpha k}$ (α = vibrational index, k = photon index) can be written in terms of an orthonormalized vibrational-field basis $|v,n\rangle \equiv |v\rangle \otimes |n\rangle$ where v runs over all unperturbed vibrational states and $|n\rangle = \exp(in\omega t)$ with n the Fourier index stepping from $-\infty$ to $+\infty$:

$$\Phi_{\alpha k}(\vec{r},t;\theta) = \exp(i\lambda_{\alpha k}t) F_{\alpha k}(\vec{r},t;\theta) \quad , \tag{5}$$

where

$$F_{\alpha k}(\vec{r},t;\theta) \equiv |\lambda_{\alpha k}\rangle_\theta = \sum_v \sum_n \exp(in\omega t)\langle vn|\lambda_{\alpha k}\rangle_\theta |v\rangle$$

$$= F_{\alpha k}\left(\vec{r},t + \frac{2\pi}{\omega} : \theta\right) \quad , \tag{6}$$

and $\lambda_{\alpha k}(\theta)$ is the (orientation dependent) quasi-vibrational energy. The QVE spectrum $\{\lambda_{\alpha k}\}$ possesses a well-defined band structure (indexed by P) characterized by the laser frequency ω (Fig. 4). All P bands (P=0,±1,±2,...) have the same OVE structure and differ from one to the other only by a phase factor. The QVEs within each P band consist of the subset $\{\lambda_{\alpha,-\alpha+p}\}$ (where $\alpha=0,1,2,...$) and are nearly degenerate and strongly coupled by the slow rotational motion. The interband rotational couplings can be shown to be orders of magnitude smaller (due to the much wider separation of QVEs) than that of intraband couplings and can be safely ignored. Thus to an excellent approximation, the IR MPA process of interest can be properly studied within any P band. For convenience let us choose the P=0 band and introduce the following short hand notations:

$$F_p \equiv F_{p,-p} \quad , \quad \lambda_p \equiv \lambda_{p,-p} \quad , \quad F_{vn,p} \equiv <vn|\lambda_{p,-p}>_\theta \quad .$$

Approximate solution of the total Hamiltonian in Eq. (3) can now be obtained by appropriate superposition of the OVE states F_p:

$$\Psi_{\epsilon_\beta}^{(m_j)}(\vec{r},t) = \exp(-i\epsilon_\beta t)(2\pi)^{-1/2} \exp(im_j\phi) \sum_p \chi_{\beta p}(\theta) F_p(\vec{r},t;\theta) \quad , \quad (7)$$

where p runs over all QVEs in the P = 0 band, and the total quasi-vibrational-rotational energies ϵ_β are to be determined. Substituting Eq. (7) into Eq. (2), averaging over a small time interval (say one period, $T = 2\pi/\omega$), and expanding $\chi_{\beta p}$ in terms of orthonormalized associated Legendre polynomials $P_j^{m_j}$:

$$\chi_{\beta p}(\theta) = \sum_{j=|m_j|} b_{\beta,pj} P_j^{m_j}(\cos\theta) \quad , \quad (8)$$

we obtain the following eigenvalue equation

$$\sum_{p'} \sum_{j'} g_{pp'}^{jj'} b_{\beta,p'j'} = \epsilon_\beta b_{\beta,pj} \quad , \quad (9)$$

where

$$g_{pp'}^{jj'} = <P_j^{m_j}|\lambda_p|P_j^{m_j}>\delta_{pp'} + \sum_v \sum_n <P_j^{m_j}|F_{vn,p}(\hat{H}_{RR}F_{vn,p'}|P_j^{m_j}>) \quad . \quad (10)$$

The structure of the total quasi-energy "G" Hamiltonian (shown in Fig. 4) is of the dimension only of $(N_V N_J)$ by $(N_V N_J)$. One thus sees that in the nonadiabatic approach, the solution of the full problem, Eq. (3), of large dimension ($N_F N_V N_J$ by $N_F N_V N_J$) is decomposed into two separate solutions,

Eqs. (4) and (9), of smaller dimensions. Finally, after making a certain unitary transformation, one arrives at the time-dependent and long-time averaged (state-to-state) MPA transition probabilities. For example, the latter is given by

$$
P_{vj \to v'j'}^{(m_j)} = \sum_m \sum_\beta \sum_n \left| \langle v'j', m-n | \epsilon_\beta^{(m_j)} \rangle \; \langle vj, -n | \epsilon_\beta^{(m_j)} \rangle \right|^2 \quad , \tag{11}
$$

where

$$
\langle vj, n | \epsilon_\beta^{(m_j)} \rangle = \sum_p \sum_{j'} b_{\beta, pj'} \langle P_j^{m_j} | F_{vn,p}(\theta) | P_{j'}^{m_j} \rangle \quad . \tag{12}
$$

The usefulness of the nonadiabatic method has been demonstrated by a detailed study of the sequential MPA spectra for the CO molecule, including state-to-state multi-quantum transitions (Fig. 5) and transitions from initially thermally distributed states as a whole.[20] Excellent agreement of the MPA spectra obtained by the nonadiabatic approach and the exact Floquet matrix method was observed in all fine details[20] (Fig. 6). Since the nonadiabatic approach is computationally order(s) of magnitude faster than the exact Floquet analysis, it may provide a practical nonperturbative method for the calculation of high resolution resonant MPA spectra of small polyatomic molecules. Work in this direction is in progress.

C. Most Probable Path Approach

The most probable path approach (MPPA) was introduced[21] recently in an <u>ab initio</u> study of high-order multiphoton excitation of the SO_2 molecule. The feasibility of multiphoton excitation (MPE) and dissociation (MPD) of triatomic molecules is a subject of experimental controversy. Some groups have reported experimental observation of collisionless MPD in triatomic molecules (SO_2, O_3, OCS) at 1 GW/cm^2 laser field strength range,[22] while others have disputed these claims.[23] It is therefore desirable to carry out comprehensive theoretical studies of the MPE/MPD dynamics of these sparse-intermediate case molecules.

A brute-force attempt to calculate polyatomic MPE would soon become unfeasible due to the large size of the Floquet matrix needed for convergence. For typical 15-photon calculations for SO_2, for example, a matrix on the order of $10,000 \times 10,000$ would have to be diagonalized at each frequency and field strength. In any exact Floquet calculation, however, the majority of the molecule-field states are unimportant due to extreme detuning or very small coupling matrix elements. The most probable path approach (MPPA) is a practical strategy introduced to determine which molecule-field states are, in fact, important at each step of the multiphoton processes. The procedure is derived from algorithms which utilize artificial intelligence to prune the number of choices at each node (photon order) of a decision tree.[24] Similar to some minimax game playing programs, the MPPA examines the possible paths to take at each photon order iteration with the static evaluation function given by Nth order perturbation theory (this is a breadth-first search). If all paths were followed exhaustively, the problem would be beyond practical solution. In game theory, one answer is to ignore paths which start with very poor moves. The MPPA likewise uses a breadth-limiting heuristic technique and discards any paths for which the Nth order coupling term is small (with respect to other Nth order terms).

The MPPA begins by calculating all possible second-order perturbative terms. The N_p largest couplings (where N_p is the number of paths to keep at each step) are chosen as the most probable paths through second order. The initial state and the intermediate states of the chosen paths are marked as important and are used in the final calculations. At each iterative step, the method calculates all possible (N+1)st-order couplings (paths) using only the N_p Nth-order paths saved in the last iteration. The (N+1)st-order couplings are then examined and the largest N_p are saved for further traversal. Nth-order states which have now become intermediate to a large (N+1)st-order path are "important" and are marked for later use. By iterating long enough, one can traverse the entire Floquet molecule-field basis space, saving only those states that are important to various ith-order processes. The reduction of the basis set is quite substantial and leads to many-orders-of-magnitude savings in computer time, yet maintains good accuracy (~95%) in most cases.

Using the MPPA, Tietz and Chu[21] found that collisionless MPD of SO_2 will not be achievable at laser intensities under 20 GW/cm^2 (Fig.

7), in agreement with the recent experimental results of Simpson and Bloembergen.[25] The latter experiment, however, has extended laser power further up to 300 GW/cm^2, and found that appreciable MPD yields begin to occur, and that the process is controlled by the laser intensity and not by the laser fluence. Further MPPA study[26] showed that the MPE of SO$_2$ is primarily a one-ladder pumping phenomenon dominated by the power broadening effect and that MPD is likely to occur (though the yield is predicted to be small, P < 10^{-3}) at laser intensities above 100 GW/cm^2.

Several other MPE theoretical studies of triatomic molecules have appeared recently. Quack and Sutcliffe[27] have studied the possibility of mode-selective IR-MPE of O$_3$, using a quasi-resonant approximation, in which they neglect interactions with states that are off resonance by more than $\hbar\omega/2$. Milfeld and Wyatt[28] have studied the MPE of OCS, using the Magnus approximation in the Floquet Hamiltonian.

It appears that detailed ab initio studies of the MPE dynamics in small polyatomic molecules are now becoming feasible. Such theoretical investigations may provide useful new physical insights and complementary information to the experimental results.

III. Many-Mode Floquet Theory for Multiphoton Processes in

Polychromatic Fields

The Floquet approaches we have described require the semiclassical Hamiltonians to be explicitly periodic in time and are therefore applicable to problems involving strictly monochromatic radiation fields. Recently there have been extensive studies of atomic and molecular processes involving the use of two lasers with different frequencies. Examples are: multiphoton double resonance experiments,[29a] collisions in two laser fields,[29b] MPD of polyatomic molecules by two IR lasers,[29c] and multiple quantum transitions in double frequency pulsed NMR experiments,[29d] etc. In addition, a broad class of various phenomena pertaining to nonlinear optics is based on experiments performed with multi-mode laser fields (for a recent review,

see Delone et al.[30]). Exact treatments of these multi-frequency problems
are beyond the scope of conventional Floquet theories.

Recently an exact extension of Shirley's (one-mode) Floquet method
to a generalized many-mode Floquet theory (MMFT) has been found.[31,32]
This makes it possible to treat the time-dependent problem of any finite-
level system exposed to polychromatic fields as an equivalent time-independent infinite-dimensional eigenvalue problem.

A. Semiclassical Many-Mode Floquet Theory

Without loss of generality, let us consider the interaction of an
arbitrary N-level quantum system with two intense linearly polarized monochromatic fields. The total Hamiltonian of the system is given by

$$\hat{H}(t) = \hat{H}_0 + \sum_{i=1}^{2} V_i(t) \tag{13}$$

where $\hat{H}_0$ is the unperturbed Hamiltonian of the N-level system,

$$\hat{H}_0|\alpha\rangle = E_\alpha|\alpha\rangle \quad , \quad \alpha = 1, 2, \ldots, N \tag{14}$$

and the interaction Hamiltonian $V_i(t)$ in the electric dipole approximation
is given by

$$V_i(t) = - \vec{\mu} \cdot \vec{\varepsilon}_i \cos(\omega_i + \phi_i) \quad , \quad i = 1,2 \tag{15}$$

with μ the electric dipole moment operator, and ε_i, ω_i and ϕ_i, respectively, the peak amplitude, frequency, and initial phase of the ith field.
According to the MMFT,[31,32] the time-dependent Schrödinger equation with
$\hat{H}(t)$ given by Eq. (13) can be transformed into an equivalent time-independent infinite-dimensional eigenvalue equation:

$$\sum_{\gamma k_1 k_2} \langle \beta n_1 n_2 | \hat{H}_F | \gamma k_1 k_2 \rangle \langle \gamma k_1 k_2 | \lambda \rangle = \lambda \langle \beta n_1 n_2 | \lambda \rangle \quad . \tag{16}$$

Here the (time-independent) Floquet Hamiltonian $\hat{H}_F$, defined in the gen-
eralized Floquet basis $\{|\beta n_1 n_2\rangle \equiv |\beta\rangle \otimes |n_1\rangle \otimes |n_2\rangle\}$, with β the system
index and n_1 and n_2 the Fourier indices, can be written as

$$\langle \beta n_1 n_2 | \hat{H}_F | \gamma k_1 k_2 \rangle = \mathcal{H}_{\beta\gamma}^{n_1 - k_1, n_2 - k_2} + (n_1 \omega_1 + n_2 \omega_2) \, \delta_{\beta\gamma} \delta_{n_1 k_1} \delta_{n_2 k_2} \qquad (17)$$

where

$$\mathcal{H}_{\beta\gamma}^{n_1 n_2} = E_\beta \delta_{\beta\gamma} \delta_{n_1 0} \delta_{n_2 0} + V_{\beta\gamma}^{(1)} (\delta_{n_1,1} \, e^{i\phi_1} + \delta_{n_1,-1} \, e^{-i\phi_1}) \, \delta_{n_2,0}$$

$$+ V_{\beta\gamma}^{(2)} (\delta_{n_2,1} \, e^{i\phi_2} + \delta_{n_2,-1} \, e^{-i\phi_2}) \, \delta_{n_1,0} \qquad (18)$$

with

$$V_{\beta\gamma}^{(i)} = - \frac{1}{2} \langle \beta | \vec{\mu} \cdot \vec{\varepsilon}_i | \gamma \rangle \quad . \qquad (19)$$

The structure of the Floquet Hamiltonian for an N-level system driven by two monochromatic linearly polarized fields is depicted in Fig. 8. Here the structure of $\hat{H}_F$ is displayed by ordering components so that β runs over the unperturbed atomic states before each change in n_1 and n_1, in turn, runs over before n_2. The quasi-energy eigenvalues $\{\lambda_{\gamma k_1 k_2}\}$ and their corresponding eigenvectors $\{|\lambda_{\gamma k_1 k_2}\rangle\}$ of $\hat{H}_F$ possess important periodic properties, namely,

$$\lambda_{\gamma, k_1 + p_1, k_2 + p_2} = \lambda_{\gamma k_1 k_2} + p_1 \omega_1 + p_2 \omega_2 \qquad (20)$$

and

$$\langle \beta, n_1 + p_1, n_2 + p_2 | \lambda_{\gamma, k_1 + p_1, k_2 + p_2} \rangle = \langle \beta n_1 n_2 | \lambda_{\gamma k_1 k_2} \rangle \quad . \qquad (21)$$

The time evolution operator, $\hat{U}(t, t_0)$, defined as the propagator that transforms the wavefunction at the initial time t_0 into the wavefunction at time t, viz.,

$$|\Psi(t)\rangle = \hat{U}(t, t_0) \, |\Psi(t_0)\rangle \quad , \qquad (22)$$

can be expressed in the following matrix form

$$U_{\beta\alpha}(t, t_0) \equiv \langle \beta | \hat{U}(t, t_0) | \alpha \rangle$$

$$= \sum_{n_1 = -\infty}^{\infty} \sum_{n_2 = -\infty}^{\infty} \langle \beta n_1 n_2 | e^{-i\hat{H}_F(t - t_0)} | \alpha 0 0 \rangle \, e^{i(n_1 \omega_1 + n_2 \omega_2)t} \quad . \qquad (23)$$

The matrix element $U_{\beta\alpha}(t,t_0)$ has the physical interpretation as the amplitude for a system initially in the generalized Floquet state $|\alpha 00\rangle$ at time t_0 propagating to the generalized Floquet state $|\beta n_1 n_2\rangle$ at time t, according to the time-independent two-mode Floquet Hamiltonian $\hat{H}_F$, summed over n_1 and n_2 with the weighting factors $\exp[i(n_1\omega_1+n_2\omega_2)t]$.

The time-dependent transition probability of going from the initial state $|\alpha\rangle$ to the final state $|\beta\rangle$ becomes

$$P'_{\alpha\to\beta}(t,t_0) = |U_{\beta\alpha}(t,t)|^2$$

$$= \sum_{m_1 m_2} \sum_{k_1 k_2} \langle\beta k_1 k_2|e^{-i\hat{H}_F(t-t_0)}|\alpha 00\rangle \cdot e^{i(m_1\omega_1+m_2\omega_2)t_0}$$

$$\cdot \langle\alpha m_1 m_2|e^{i\hat{H}_F(t-t_0)}|\beta k_1 k_2\rangle \quad . \tag{24}$$

Performing the averaging over random initial times t_0 and then the averaging over all possible elapsed times $t-t_0$, one obtains the long-time averaged transition probability

$$\bar{P}_{\alpha\to\beta} = \sum_{k_1 k_2} \sum_{\gamma\ell_1\ell_2} |\langle\beta k_1 k_2|\lambda_{\gamma\ell_1\ell_2}\rangle \cdot \langle\lambda_{\gamma\ell_1\ell_2}|\alpha 00\rangle|^2 \quad . \tag{25}$$

The many-mode Floquet theory outlined above has been applied to the study of dynamical evolution of multi-quantum NMR transitions in spin-1/2 systems[33] and to resonance lineshape analysis in three-level systems[34] subjected to intense bichromatic fields. In addition, we have extended the almost degenerate perturbation theory of Salwen[35] to the two-mode Floquet Hamiltonian and have obtained analytical formulas for the first time for multiphoton transition probabilities, resonance bichromatic (Bloch-Siegert) shifts and widths, and absorption lineshapes, beyond the conventional rotating wave approximation (RWA). Some of the analytical results are outlined in the Appendix.

B. SU(N) Dynamical Symmetries and Nonlinear Coherence

Multiphoton excitation (MPE) of N-level atomic and molecular systems driven by more than one laser field is a subject of intense current

experimental and theoretical interest.[29,30,36] In particular, the dynamical evolution of three-level systems in the presence of two strong monochromatic fields has been central to discussions of two-photon NMR on the
Spin-1 system,[37] two-photon coherence,[38] coherent population trapping,[39]
resonant Raman scattering and double-resonance processes,[40] trilevel
echoes,[41] and many others.

It has long been known that for two-level systems, the description
of magnetic and optical resonance phenomena can be greatly simplified by
the use of the Bloch spin or pseudospin vector.[42] However, extension of
the vector description to more complex systems has not been achieved until
recently. Hioe and Eberly[43] and others found that the dynamical evolution
of $N(\geq 3)$-level systems can be expressed in terms of the generalized rotation of an (N^2-1) dimensional real coherence vector $\vec{S}$ whose property can
be analyzed by appealing to SU(N) group symmetry. For example, the time
evolution of three-level systems can be described by a coherent vector of
constant length rotating in an eight-dimensional space.[44] The existence
of a number of unexpected nonlinear constants of motion that govern the
density matrix of an N-level system has also been noticed. In particular,
for a three-level system under the two-photon resonance condition, the
time-evolution of the eight-dimensional coherent vector $\vec{S}$ can be analyzed
in terms of the time evolution of three independent vectors of dimensions
three, four, and one, rotating in three disjoint subspaces of those dimensions, provided that the rotating wave approximation (RWA) is valid. The
three nonlinear constants of motion in this case correspond to the squares
of the lengths of these three subvectors. The dynamical symmetry underlying the three-level system is reminiscent of the Gell-Mann SU(3) symmetry[45]
in particle physics. Thus the subspaces of three-, four-, and one-dimen-
sion of $\vec{S}$ are analogous to the subspaces of pions (π^+, π^0, π^-), kaons (K^+,
K^0 $\bar{K}^+$, $\bar{K}^0$), and eon (n^0), respectively. In practice, however, if the
laser-atom interactions occur away from the two-photon resonance, or if
the RWA is not valid, or if decays are taken into account, then the dy-
namical subspaces ($\underline{8} = \underline{3} \oplus \underline{4} \oplus \underline{1}$) discussed by Hioe and Eberly will
no longer be valid. The Gell-Mann SU(3) symmetry[45] of the system will
then be broken.

The study of SU(N) dynamical evolution of the coherent vector $\vec{s}$ and the symmetry-breaking effects embodied in N-level systems subjected to an arbitrary number of monochromatic fields can be greatly facilitated by the use of MMFT.[46] In the following we confine our discussion to the SU(3) case corresponding to the dynamical evolution of a three-level system driven by a bichromatic field. We adopt the standard form of the SU(3) generators used by Gell-Mann[45] (which are not restricted to the RWA), namely,

$$\hat{s} \equiv \{\hat{s}_i \mid i=1,\ldots 8\}$$

where

$$\hat{s}_1 = \begin{pmatrix} 0 & 1 & 0 \\ 1 & 0 & 0 \\ 0 & 0 & 0 \end{pmatrix} \;,\quad \hat{s}_2 = \begin{pmatrix} 0 & -i & 0 \\ i & 0 & 0 \\ 0 & 0 & 0 \end{pmatrix} \;,\quad \hat{s}_3 = \begin{pmatrix} 1 & 0 & 0 \\ 0 & -1 & 0 \\ 0 & 0 & 0 \end{pmatrix} \;,$$

$$\hat{s}_4 = \begin{pmatrix} 0 & 0 & 1 \\ 0 & 0 & 0 \\ 1 & 0 & 0 \end{pmatrix} \;,\quad \hat{s}_5 = \begin{pmatrix} 0 & 0 & -i \\ 0 & 0 & 0 \\ i & 0 & 0 \end{pmatrix} \;,\quad \hat{s}_6 = \begin{pmatrix} 0 & 0 & 0 \\ 0 & 0 & 1 \\ 0 & 1 & 0 \end{pmatrix} \;,$$

$$\hat{s}_7 = \begin{pmatrix} 0 & 0 & 0 \\ 0 & 0 & -i \\ 0 & i & 0 \end{pmatrix} \;,\quad \hat{s}_8 = \frac{1}{\sqrt{3}} \begin{pmatrix} 1 & 0 & 0 \\ 0 & 1 & 0 \\ 0 & 0 & -2 \end{pmatrix} \;. \tag{26}$$

In terms of the eight 3×3 matrices in Eq. (26) and the 3×3 identity matrix $\hat{I}$, the Hamiltonian for a three-level system in a bichromatic field in Eq. (13) can be expressed as

$$\hat{H}(t) = \frac{1}{3} \left(\sum_{\alpha=1}^{3} E_\alpha \right) \hat{I} + \frac{1}{2} \sum_{j=1}^{8} \Gamma_j(t) \hat{s}_j \;, \tag{27}$$

and the density matrix of the system can be written as

$$\hat{\rho}(t) = \frac{1}{3} \hat{I} + \frac{1}{2} \sum_{j=1}^{8} S_j(t) \hat{s}_j \tag{28}$$

where

$$\Gamma_j(t) = \text{Trace} \, [\hat{H}(t)\hat{s}_j] \;, \tag{29}$$

and

$$S_j(t) = \text{Trace}[\hat{\rho}(t)\,\hat{s}_j] \;. \tag{30}$$

Substituting Eqs. (27) and (28) into the Louville equation

$$i \frac{\partial \hat{\rho}(t)}{\partial t} = [\hat{A}(t), \hat{\rho}(t)] \tag{31}$$

results in an equation of motion (i.e. the generalized Bloch equation) for the coherence vector $\vec{S}(t)$, namely

$$\frac{d}{dt} \vec{S}_j = \sum_{k\ell} f_{jk\ell} \Gamma_k(t) \vec{S}_\ell \tag{32}$$

where $f_{jk\ell}$'s are the structure constants associated with the Gell-Mann type generators, i.e., Eqs. (26) of the group SU(3). The length of the coherent vector $\vec{S}(t)$ is a constant of motion, and is given by $K \equiv |\vec{S}|^2 = 2\{Tr(\hat{\rho}\hat{\rho}) - \frac{1}{3}\}$.

At the exact two-photon resonance condition (i.e. $\Delta_1 + \Delta_2 = 0$, where Δ_1 and Δ_2 are the detunings which are given by, for a cascade system ($E_1 < E_2 < E_3$), $\Delta_1 = (E_2-E_1) - \omega_1$ and $\Delta_2 = (E_3-E_2) - \omega_2$), within the RWA, $\vec{S}(t)$ can be factored into three subvectors, namely,

$$\vec{S}(t) = \vec{A}(t) + \vec{B}(t) + \vec{C}(t) \tag{33}$$

where $\vec{A}(t)$, $\vec{B}(t)$, and $\vec{C}(t)$, of dimensions three, four, and one, rotate independently, and their respective lengths are preserved in the course of time. Assuming that the system is a cascade type (Fig. 9) and is in its ground state initially, and that the initial phases θ_1 and θ_2, of the bichromatic fields, are zero, we obtain analytical expressions for the components of $\vec{A}(t)$, $\vec{B}(t)$, and $\vec{C}(t)$ as

$$A_1(t) = - \frac{2\Omega_1^2 \Delta}{b \cdot p^2} (1-\cos pt) \quad , \tag{34a}$$

$$A_2(t) = \frac{2\Omega_1^2}{b \cdot p} \sin pt \quad , \tag{34b}$$

$$A_3(t) = - \frac{4\Omega_1^2}{p^2} (1-\cos pt) + \frac{\Omega_1^2}{b^2} \quad , \tag{34c}$$

$$B_1(t) = -\frac{2\Omega_1\Omega_2'}{p} \left(\frac{1}{\xi}\sin\xi t - \frac{1}{\eta}\sin\eta t\right) \quad , \tag{35a}$$

$$B_2(t) = \frac{2\Omega_1\Omega_2'}{b \cdot p}(\cos\xi t - \cos\eta t) \quad , \tag{35b}$$

$$B_3(t) = \frac{2\Omega_1 \cdot \Omega_2'}{b\Delta \cdot p}\left\{-\xi\left(1-\frac{b^2}{\xi^2}\right)\sin\xi t + \eta\left(1-\frac{b^2}{\eta^2}\right)\sin\eta t\right\} \quad , \tag{35c}$$

$$B_4(t) = \frac{2\Omega_1\Omega_2'}{b^2} + \frac{2\Omega_1\Omega_2'}{\Delta \cdot p}\left\{-\left(1-\frac{b^2}{\xi^2}\right)(1-\cos\xi t) + \left(1-\frac{b^2}{\eta^2}\right)(1-\cos\eta t)\right\} \quad , \tag{35d}$$

and

$$C(t) = \frac{2\Omega_2'^2 - \Omega_1^2}{\sqrt{3}\, b^2} \equiv \sqrt{K_3} \quad , \tag{36}$$

where

$$\Omega_1 = \frac{1}{2}\langle 1|\vec{\mu}\cdot\vec{\varepsilon}_1|2\rangle \quad , \quad \Omega_2' = \frac{1}{2}\langle 2|\vec{\mu}\cdot\vec{\varepsilon}_2|3\rangle \quad , \quad b = (\Omega_1^2 + \Omega_2'^2)^{1/2} \quad ,$$

$$\Delta = \Delta_1 = -\Delta_2 \neq 0 \quad ,$$

$$p = (\Delta^2 + 4b^2)^{1/2} \quad ,$$

and

$$\xi = \frac{1}{2}p + \frac{\Delta}{2} \quad ,$$

and

$$\eta = \frac{1}{2}p - \frac{\Delta}{2} \quad .$$

From Eqs. (34) and (35), and the conservation of the lengths of vectors $\vec{A}(t)$ and $\vec{B}(t)$, we can see that the vector $\vec{A}(t)$ pursues a circular motion defined by the relations

$$\frac{b^2 p^2}{4\Omega_1^4}\left\{\frac{p^2}{\Delta^2}\left(A_1(t) + \frac{2\Omega_1^2 \cdot \Delta}{b \cdot p^2}\right)^2 + A_2(t)^2\right\} = 1 \quad , \tag{37a}$$

and

$$A_1(t)^2 + A_2(t)^2 + A_3(t)^2 = \frac{\Omega_1^4}{b^4} \equiv K_1 \quad , \tag{37b}$$

while the vector $\vec{B}(t)$ engages in a much more complicated motion defined by the relations

$$B_2(t)^2 + B_3(t)^2 = \frac{8\Omega_1^2\Omega_2'^2}{b^2p^2} \{1 - \cos pt\} \quad , \tag{37c}$$

$$B_1(t)^2 + B_4(t)^2 = \frac{4\Omega_1^2\Omega_2'^2}{b^4} \{1 - \frac{2b^2}{p^2} (1 - \cos pt)\} \quad , \tag{37d}$$

and

$$B_1(t)^2 + B_2(t)^2 + B_3(t)^2 + B_4(t)^2 = \frac{4\Omega_1^2\Omega_2'^2}{b^4} \equiv K_2 \quad . \tag{37e}$$

It is thus clear that the projection of the trajectory of $\vec{A}(t)$ on either the A_1-A_2, or the A_1-A_3 plane is an ellipse, whereas projections of the trajectory of $\vec{B}(t)$ on B_2-B_3 and B_1-B_4 planes, respectively, are spinning cardioid and cycloid characterized by the frequency p. Note that at the exact two-photon resonance $\Delta_1 = -\Delta_2 = \Delta$, but $\Delta \neq 0$, in the RWA limit, the three relevant quasi-energies become 0, ξ, and $-\eta$, respectively. The quantity p is simply the difference between the quasi-energies $\lambda_{2,-1,0}$ and λ_{100} (or $\lambda_{3,-1,-1}$), and is predominantly of one-photon character. Therefore, we can see that the sub-coherence vector $\vec{A}(t)$ depicts the one-photon coherence between the quasi-energy states $|\lambda_{2,-1,0}\rangle$ and $|\lambda_{100}\rangle$ (or $|\lambda_{3,-1,-1}\rangle$), while the vector $\vec{B}(t)$ describes the two-photon coherences between the remaining pairs of quasi-energy states $|\lambda_{2,-1,0}\rangle$ and $|\lambda_{3,-1,-1}\rangle$ (or $|\lambda_{100}\rangle$), and $|\lambda_{100}\rangle$ and $|\lambda_{3,-1,-1}\rangle$.

In more general (dissipationless) cases, deviations from either the RWA limit, or the two-photon resonance condition, will modify the trajectory of the $\vec{S}(t)$ described by Eqs. (34) via (36), and thus break the dynamic symmetries embodied in the independence of the subvectors $\vec{A}$, $\vec{B}$, and $\vec{C}$ in the course of the time. The study of this symmetry breaking effect can be facilitated by means of Eq. (30) and the relation

$$\hat{\rho}(t) = U(t,t_0) \; \hat{\rho}(t_0) \; U^+(t,t_0) \quad . \tag{38}$$

Here $\hat{\rho}(t_0)$ is the density matrix at the initial time t_0 (initial conditions) and the time-evolution operator $U(t,t_0)$ can be determined by the method of MMFT, Eq. (23), and expressed in terms of a few time-independent quasi-energy eigenvalues and eigenvectors. Furthermore, the generalized Van Vleck (GVV) nearly degenerate perturbation theory[47] can be extended to the analytical treatment of the time-independent many-mode Floquet Hamiltonian. The general idea behind the MMFT-GVV technique is to block-diagonalize the time-independent Floquet Hamiltonian $\hat{H}_F$ (such as Fig. 8) so that the coupling between the model space (consisting of nearly degenerate and strongly coupled Floquet states of interest) and the remainder of the configuration space (called the external space) diminishes to a desired order. One important feature of the MMFT-GVV approach is that if the perturbed model space wave functions are exact to the nth order, the corresponding quasi-energy eigenvalues in the model space will be accurate to the (2n+1)th order. In that regard, it is interesting to note that the RWA is merely the lowest order (i.e., n = 0) limit, namely, model space wave functions correct only to the zeroth order and eigenvalues accurate to the first order. Furthermore, while the RWA can only deal with sequential one-photon processes, the MMFT-GVV approach is capable of treating both one-photon and multiphoton processes on an equal footing. Thus the MMFT-GVV approach appears to be a natural and powerful extension beyond the conventional RWA limit for the nonperturbative treatment of multiphoton processes in intense polychromatic fields.

The MMFT-GVV method has been recently applied to a unified treatment of both the SU(3) symmetries (RWA limit) and symmetry-breaking effects (caused by non-RWA terms) of dissipationless three-level systems at two-photon resonances induced by intense bichromatic fields.[46] The MMFT-GVV technique reduces the infinite-dimensional time-independent two-mode Floquet Hamiltonian to a 3×3 (model space) effective Hamiltonian, from which essential analytical properties and vivid geometry of the eight-dimensional coherence vector are revealed for the first time. Figures 10-16 illustrate the general features of the dynamical evolution of a cascade three-level system ($E_1 < E_2 < E_3$, $(E_2-E_1) \gg (E_3-E_2)$) at two-photon resonance condition (i.e. detunings $\Delta_1 = -\Delta_2 = \Delta$) driven by two linearly polarized monochromatic fields (Fig. 9a). Figure 10 shows the time-dependent transition probabilities $P_{1 \to \beta}(t)$, $\beta = 1,2,3$, while Fig. 11(12) shows the components of the three- (four-) dimensional coherent subvector $\vec{A}(t)$ ($\vec{B}(t)$). The projections of the eight-dimensional coherent vector $\vec{S}(t)$ trajectory onto the (A_1-A_2), (B_1-B_4), and (B_2-B_3) planes are depicted in Figs. 13, 14, and 15 respectively. Here the RWA results (SU(3) symmetry) are obtained by Eqs. (34) and (35), while the symmetry-breaking results are determined by the MMFT-GVV procedure. Finally, in Fig. 16, we sketch the lengths of the three subvectors $\vec{A}(t)$, $\vec{B}(t)$, and $\vec{C}(t)$, and of the coherence vector $\vec{S}(t)$ as a function of time. The dotted lines show constancies of the subvectors in the RWA limit, while the solid curves show the corresponding GVV symmetry-breaking results. In this case the coupling between the vectors $\vec{A}(t)$ and $\vec{C}(t)$ appears much stronger than their individual coupling with the vector $\vec{B}(t)$, though the length of the vector $\vec{B}(t)$ also exhibits small oscillations away from the RWA constant. The length of the coherence $\vec{S}(t)$ stays constant in either calculation, and assumes the value $K = \left| \vec{S}(t) \right|^2 = 4/3$, as it should. Other case studies can be found in Ref. 46.

More recently, the MMFT has been extended to incorporate the irreversible damping mechanisms (described by phenomenological damping constants) for the nonperturbative treatment of the dynamical evolution of <u>dissipative</u> three-level systems at two- or multi-photon coherent resonance trapping conditions induced by two strong linearly polarized monochromatic fields.[48] It has been recently shown by several workers[49] that under the rotating wave approximation, population may be permanently trapped in the three-level system if the coherent monochromatic fields are exactly two-photon resonant with the initial and final states, decoupled from the intermediate decaying level. The deep "dark" resonances[39,50] or the narrow coherence minima observed in the Raman absorption lineshape or the total fluorescence signal as a function of detuning are thought to be a manifestation of the coherent trapping of the population. These narrow trapping holes have been extensively studied with a view to possible applications in high-resolution spectroscopy and for potential time and frequency standards.[51] The effects of finite field bandwidths, spontaneous emissions and/or correlation between fields on coherent population trapping have also been studied and it has been found that the trapping can be either destroyed, preserved or even enhanced.[49] What is not known still is the effects of non-RWA terms on the trapping, that is, how and to what extent the coherent trapping conditions will be modified when the RWA model breaks down in intense fields. This problem is taken up in Ref. 48.

We have extended the generalized Van Vleck (GVV) nearly degenerate perturbation theory to an analytical treatment of the non-Hermitian two-mode Floquet Hamiltonian. This reduces the infinite-dimensional time-independent non-Hermitian Floquet Hamiltonian to a 3×3 effective Hamiltonian, from which essential properties of the coherent population trapping behavior as well as the dynamical evolution of the dissipative SU(3) coherence vector $\vec{S}(t)$ can be readily obtained and expressed in terms of only three complex quasi-energy eigenvalues and eigenvectors. The MMFT-GVV studies show that the RWA two-photon resonant trapping condition is substantially modified by the effects of non-RWA terms, and that the system can be "quasi-trapped" for only a finite amount of time characterized by a small imaginary energy (width) associated with a coherent superposition state of the initial and final levels. Furthermore, it is found that the initially eight-dimensional coherence vector $\vec{S}(t)$ evolves predominantly to a one-dimensional scalar at the two- or multiphoton resonance quasi-trapping conditions. Detailed results and pictorial representations of the population trapping and SU(3) dissipative dynamical evolution are described in Ref. 48.

IV. Multi-Charged Ion-Atom Collisions in Laser Fields: Coupled Dressed-Quasi-Molecular-States (DQMS) Approaches

Information about the charge-exchange cross sections in multi-charged ion-atom collisions plays an essential role in the interpretation and understanding of many physical phenomena such as determining the radiation losses and neutral beam heating efficiencies in Tokamak plasmas[52] and predicting the feasibility of producing an X-ray laser.[53]

It has been found that the nonresonant ion-atom charge transfer cross section, which is normally small at low velocities, can be greatly enhanced by the presence of a strong laser field.[54] The observation of a laser-induced charge exchange collision involving Ca^+ and Sr has also been reported.[55] More recently, Seely and Elton[56] have suggested that photon-induced charge transfer reactions are also useful for the measurement of particle densities, with time and space resolution provided by the pumping laser beam, in a Tokamak plasma.

Theoretical studies[57-63] have shown that at weak laser fields and at small impact velocities, the laser assisted processes $A^+ + B + \hbar\omega \to A + B^+$ can be described by perturbative schemes of one kind or another within the rotating wave approximation (RWA) and the impact parameter formalism, and that the charge transfer cross sections behave linearly as a function of the intensity of the laser field and as a function of the reciprocal of the impact velocity. These approaches, however, are generally difficult to extend to high laser intensity and very low collision velocity regions where the laser-induced charge transfer rates tend to be the largest. Recently we have developed a nonperturbative coupled dressed-quasimolecular-states (DQMS) formalism[64] for the treatment of charge transfer processes at low collision velocities and strong laser intensities. The theory parallels an earlier treatment[65] in the study of multiphoton enhancement of vibrational excitations induced by molecular collisions. The essence of the coupled DQMS approach is as follows: As the laser frequency of interest is in the range of quasimolecular electronic energy separations, the laser field oscillates more rapidly than the nuclear motion. It is legitimate then to first construct the solutions of the $(A-B)^+$ + field, system, namely, the dressed quasimolecular electronic states (also called the quasi-energy or electronic-field states) with the internuclear separation R fixed. The laser assisted collision processes can then be treated as the electronic transitions among the DQMS driven by the nuclear motion only. The DQMS can be determined by invoking either the full quantized treatment[66] or the semiclassical Floquet theory,[16,17] both treatments being equivalent in strong fields. In our formalism,[64] we have adopted the Floquet approach. The essence of the coupled DQMS approaches is described in Secs. IVA, IVB, IVC.

A. Adiabatic DQMS method

Within the semiclassical trajectory formulation, the slow ion-atom collisions in the presence of an intense linearly polarized laser field can be described (in the electric dipole approximation) by the time-dependent Schrödinger equation

$$i \frac{\partial}{\partial t} \Psi(\vec{r},t|\vec{R}(t)) = \hat{H}(\vec{r},t|\vec{R}(t))\Psi(\vec{r},t|\vec{R}(t)) \quad , \tag{39}$$

where $\Psi(\vec{r},t|\vec{R}(t))$ is the state wave function of the colliding system at some instant of the time t, $\hat{H}(\vec{r},t|\vec{R}(t))$, the total Hamiltonian, consists of a field-free electronic Hamiltonian $\hat{h}_{e\ell}(\vec{r}|\vec{R}(t))$ and an interaction $\hat{V}(\vec{r},t) = -\vec{r}\cdot\vec{E}_0 \cos\omega t$ between the active electron, positioned at $\vec{r}$, and the applied laser field of frequency ω and amplitude $\vec{E}_0(\theta,\phi)$. The $\hat{z}$-direction is chosen parallel to the initial impact velocity $\vec{v}_0$ of the two nuclei. The nuclear trajectory $\vec{R}(t)$ is determined by some average potential of the two colliding entities, and for simplicity will be taken as rectilinear, as in the impact parameter method. Therefore, for each specified impact parameter $\vec{b} \perp \vec{v}_0$, we have $\vec{R}(t) = \vec{b} + \vec{v}_0 t$. If we assume that the frequency of the laser field is of the same order of magnitude as that characterizing the electronic motion, i.e. much greater than that characterizing the nuclear motion, the electronic states can be first thought of as dressed by the applied laser field before adjusting to the changing molecular field caused by the nuclear motion. The adiabatic dressed-quasi-molecular state (DQMS) wave function $\phi_{\beta n}^a(\vec{r},t|\vec{R}(t))$ obeys the eigenvalue equation obtained by the Floquet theorem[64]

$$\{\hat{H} - i(\frac{\partial}{\partial t})_{\vec{R}}\}\phi_{\beta n}^a(\vec{r},t|\vec{R}(t)) = \varepsilon_{\beta n}^a(R(t))\phi_{\beta n}^a(\vec{r},t|\vec{R}(t)) \tag{40}$$

defined in an extended Hilbert space $R \oplus T$ with the spatial part R spanned by the eigenfunctions $\psi_\beta(\vec{r}|\vec{R}(t))$ of the field-free electronic Schrödinger equation

$$\hat{h}_{e\ell}(\vec{r}|\vec{R})\psi_\beta = E_\beta(R)\psi_\beta \quad , \quad \beta = 1,2,\dots N \quad , \tag{41}$$

and the temporal part T by the complete orthonormal set of functions $e^{in\omega t}$, $n = 0, \pm 1, \pm 2, \dots \pm\infty$, which obey the relation

$$\frac{\omega}{2\pi} \int_0^{2\pi/\omega} e^{i(m-n)\omega t} \, dt = \delta_{mn} \quad . \tag{42}$$

The time derivative $(\partial/\partial t)_{\vec{R}}$ in Eq. (40) is taken at some fixed internuclear distance $\vec{R}$. The DQMS potential energies $\varepsilon_{\beta n}^a(R(t))$ thus adjust continuously to the change of the internuclear separation R. By expanding the total wave function $\Psi(\vec{r},t|\vec{R}(t))$ of Eq. (39) in terms of the DQMS wave functions $\phi_{\beta n}^a(\vec{r},t|\vec{R}(t))$, i.e.

$$\Psi(\vec{r},t|\vec{R}(t)) = \sum_{\beta n} \chi_{\beta n}^a(\vec{R}(t))\phi_{\beta n}^a(\vec{r},t|\vec{R}(t)) \tag{43}$$

and inserting it into Eq. (39), we obtain a set of coupled <u>adiabatic</u> DQMS equations for $\chi_{\beta n}^a(\vec{R}(t))$, namely,

$$i\,\frac{\partial}{\partial t}\,\chi_{\beta n}^a(\vec{R}(t)) = \varepsilon_{\beta n}^a(R(t))\chi_{\beta n}^a(\vec{R}(t)) - i\dot{\vec{R}} \cdot \sum_{\alpha m} \hat{A}_{\beta n,\alpha m}^a(\vec{R}(t))\chi_{\alpha m}^a(\vec{R}(t)) \quad , \tag{44}$$

where the nonadiabatic coupling matrix element $\hat{A}_{\beta n,\alpha m}^a(\vec{R}(t))$ between two arbitrary DQMS wave functions $\phi_{\beta n}^a(\vec{r},t|\vec{R}(t))$ and $\phi_{\alpha m}^a(\vec{r},t|\vec{R}(t))$ can be written as

$$\hat{A}_{\beta n,\alpha m}^a(\vec{R}(t)) \equiv \langle\langle \phi_{\beta n}^a | \nabla_{\vec{R}} | \phi_{\alpha m}^a \rangle\rangle_R$$

$$= \frac{\omega}{2\pi} \int_0^{2\pi/\omega} dt \int d\vec{r} \, \phi_{\beta n}^a \, \nabla_{\vec{R}} \, \phi_{\alpha m}^a \quad . \tag{45}$$

Equation (40) can be solved by expanding the DQMS wave function $\phi_{\beta n}^a(\vec{r},t|\vec{R}(t))$, which is a periodic function of t with period $2\pi/\omega$ at each fixed internuclear separation R, in terms of the field-free quasimolecular-state wave functions $\{\psi_\alpha\}$ and the Fourier basis $\{e^{im\omega t}\}$,

$$\phi_{\beta n}^a(\vec{r},t|\vec{R}) = \sum_{\alpha m} \langle \alpha m | \varepsilon_{\beta n}^a \rangle e^{im\omega t}\psi_\alpha(\vec{r},\vec{R}) \quad . \tag{46}$$

Here the expansion coefficient $\langle \alpha m | \varepsilon^a_{\beta n} \rangle$ satisfies an infinite-dimensional eigenvalue equation

$$\sum_\gamma \sum_k \langle \alpha m | \hat{H}_F | \gamma k \rangle \langle \gamma k | \varepsilon^a_{\beta n} \rangle = \varepsilon^a_{\beta n}(R) \ \langle \alpha m | \varepsilon^a_{\beta n} \rangle \tag{47}$$

with the Floquet Hamiltonian $\hat{H}_F$ at each fixed R defined as

$$\langle \alpha m | \hat{H}_F | \gamma k \rangle (R) = E_\alpha(R) \delta_{\alpha\gamma} \delta_{mk} + m\omega \delta_{\alpha\gamma} \delta_{mk} - \frac{\vec{E}_0}{2} \cdot \langle \psi_\alpha | \vec{r} | \psi_\gamma \rangle \cdot$$
$$\cdot (\delta_{m,k+1} + \delta_{m,k-1}) \quad . \tag{48}$$

In terms of coefficients $\{ \langle \alpha m | \varepsilon^a_{\beta n} \rangle (R) \}$ the nonadiabatic coupling matrix elements $\hat{A}^a_{\beta n, \alpha m}(\vec{R})$ can be written as

$$\hat{A}^a_{\beta n, \alpha m}(\vec{R}) = \hat{B}^a_{\beta n, \alpha m}(\vec{R}) + \hat{D}^a_{\beta n, \alpha m}(\vec{R}) \tag{49}$$

where $\hat{B}$ is the field-dominated coupling,

$$\hat{B}^a_{\beta n, \alpha m}(\vec{R}) = \frac{\vec{R}}{R} \sum_\gamma \sum_k \langle \varepsilon^a_{\beta n} | \gamma k \rangle \frac{d}{dR} \langle \gamma k | \varepsilon^a_{\alpha m} \rangle \quad , \tag{50}$$

and $\hat{D}$ is the collision-dominated coupling

$$\hat{D}^a_{\beta n, \alpha m}(\vec{R}) = \sum_{\gamma_2} \sum_{\gamma_1} \sum_k \langle \varepsilon^a_{\beta n} | \gamma_2 k \rangle \langle \gamma_1 k | \varepsilon^a_{\alpha m} \rangle \langle \psi_{\gamma_2} | \nabla_{\vec{R}} | \psi_{\gamma_1} \rangle \quad . \tag{51}$$

The field-dominated coupling $\hat{B}^a_{\beta n, \alpha m}(\vec{R})$ becomes important when the two field-free electronic states, denoted by $E_\beta(R)$ and $E_\alpha(R)$, are strongly mixed by the applied laser field. The collision-dominated coupling $\hat{D}^a_{\beta n, \alpha m}(\vec{R})$ becomes significant only when the corresponding field-free electronic potential energies $E_\beta(R)$ and $E_\alpha(R)$ possess pseudo-crossings, at some internuclear separation R. In the adiabatic DQMS representation, cf. Eq. (44), states which are strongly coupled are thus indicated by regions of avoided crossings, a manifestation of the size of either $\hat{B}^a_{\beta n, \alpha m}(\vec{R})$, or $\hat{D}^a_{\beta n, \alpha m}(\vec{R})$, or both, in the correlation diagram of the DQMS potential energies $\varepsilon^a_{\beta n}(R)$. Therefore, in the coupled adiabatic DQMS equations (44), the collision-induced and the field-induced transition mechanisms are treated on an equal footing. However, in the region of low collisional velocity and high laser intensity, one can easily show that the collision-dominated coupling $\hat{D}^a$ is negligibly small compared with the field-dominated coupling $\hat{\beta}^a$.

Equation (44), our working equation, is to be solved subject to the initial conditions

$$\chi^a_{\alpha m}(R) \rightarrow \delta_{\alpha\mu}\delta_{mo} \quad , \quad \text{at} \quad t = -\infty$$

which implies that

$$\Psi(\vec{r},t|R) \rightarrow \phi_{\mu 0}(\vec{r},t|R) \rightarrow \psi_\mu(\vec{r}|R) \quad , \quad \text{at} \quad t = -\infty$$

with the index μ specifying the initial state of the system before entering the laser field. It is interesting to see that

$$<\beta n|\epsilon_{\alpha m}> \rightarrow \delta_{\alpha\beta}\delta_{mn} \quad , \quad \text{as} \quad R \rightarrow \infty \tag{52}$$

and, therefore

$$\Psi(\vec{r},t|R) \rightarrow \sum_\alpha \sum_{m=-\infty}^{\infty} (\chi_{\alpha m}e^{im\omega t})\psi_\alpha(\vec{r}|R) \quad , \quad \text{as} \quad R \rightarrow \infty \quad . \tag{53}$$

Equations (52) and (53) indicate that the probability of finding the system in a field-free quasimolecular state ν after the collision is

$$P_{\mu\nu}(\rho,\vec{V}_0;\vec{E}_0,\omega) = \sum_{m=-\infty}^{\infty} |\chi_{\nu m}(t=+\infty)|^2 \tag{54}$$

for each specified impact parameter ρ and initial nuclear impact velocity $\vec{V}_0$. The total cross section for the system going from the initial state μ to the final state ν can thus be written as

$$\sigma_{\mu\nu}(\vec{V}_0,\vec{E}_0,\omega) = 2\pi \int_0^\infty d\rho \ \rho P_{\mu\nu}(\rho,\vec{V}_0;\vec{E}_0,\omega) \quad . \tag{55}$$

It is instructive to examine the coupled adiabatic Eq. (44) in the limit of $|\vec{E}_0|$ approaching zero. Noticing that

$$\epsilon^a_{\alpha m}(R) \rightarrow E_\alpha(R) + m\omega$$

and

$$<\beta n|\epsilon^a_{\alpha m}> \rightarrow \delta_{\alpha\beta}\delta_{mn} \quad , \quad \text{as} \quad |\vec{E}_0| \rightarrow 0 \quad , \tag{56}$$

we find that Eq. (44) reduces to

$$i \frac{\partial}{\partial t} \chi^a_{\beta n}(R) = E_\beta \chi^a_{\beta n}(R) - i\dot{\vec{R}} \cdot \sum_\alpha <\psi_\beta |\nabla_{\vec{R}}| \psi_\alpha> \chi^a_{\alpha n}(R) \quad , \tag{57}$$

which represents simply the coupled equations for the case of field-free collisions if one sets n equal to zero.

B.   Diabatic DQMS method

The DQMS defined by Eq. (40) form a set of adiabatic bases which incorporate both the motion of the nuclei and the presence of the laser field. The corresponding adiabatic DQMS energy curves $\varepsilon_{am}(R)$ will have regions of avoided crossings which, in turn, provoke strong couplings, represented by the matrix elements of the operator $\nabla_{\vec{R}}$ in Eq. (44), between the adiabatic DQMS. In the regions of avoided crossings of the adiabatic DQMS energy curves, a more appropriate physical description may be obtained by transforming the adiabatic DQMS bases into a set of diabatic DQMS bases for which the coupling matrix elements of $\nabla_{\vec{R}}$ are small and transitions are driven by a potential matrix. Following closely the definition of the diabatic states introduced by Smith[67a] and Heil and Dalgarno,[67b] we can construct, from the adiabatic DQMS basis $\{\phi^a_{\alpha m}(r,t|R)\}$ defined by Eq. (40), the diabatic DQMS basis $\{\phi^d_{\alpha m}\}$ via the transformation

$$\phi^d_{\beta n}(R) = \sum_\alpha \sum_m \phi^a_{\alpha m} \hat{C}_{\alpha m, \beta n}(R) \quad , \tag{58}$$

where the matrix $[C_{\alpha m, \beta n}(R)]$ satisfies the relation

$$\sum_\gamma \sum_k [\hat{A}^a_{\alpha m, \gamma k}(\vec{R}) \hat{C}_{\gamma k, \beta n}(R)] + \nabla_{\vec{R}} \hat{C}_{\alpha m, \beta n}(R) = 0 \tag{59}$$

with $C_{\gamma k, \beta n}(R) \rightarrow \delta_{\gamma \beta} \delta_{kn}$ at large R, and

$$\hat{A}^a_{\alpha m, \gamma k}(\vec{R}) \equiv <<\phi^a_{\alpha m}|\dot{\nabla}_{\vec{R}}|\phi^a_{\gamma k}>>_R \quad . \tag{60}$$

Equations (58) and (59) ensure that

$$<<\phi^d_{\alpha m}|\nabla_{\vec{R}}|\phi^d_{\beta n}>>_R = 0 \tag{61}$$

and we obtain a set of the coupled <u>diabatic</u> equations as follows:

$$i \frac{\partial}{\partial t} x_{\beta n}^{d}(\vec{R}) = \sum_{\alpha} \sum_{m} V_{\beta n,\alpha m}[\vec{R}(t)]x_{\alpha m}^{d}[\vec{R}(t)] \quad , \tag{62}$$

where the potential matrix [V] is defined by the relation

$$V_{\beta n,\alpha m} = \sum_{\gamma} \sum_{k} (C^{-1})_{\beta n,\gamma k}(R)\epsilon_{\gamma k}^{a}(R)C_{\gamma k,\alpha m}(R) \quad . \tag{63}$$

Equation (62) is then solved subject to the initial conditions

$$x_{\alpha m}^{d}(R) = \delta_{\alpha\mu}\delta_{mo} \quad \text{at} \quad t = -\infty \tag{64}$$

with μ indexing the initial state of the colliding system. The transition probability and the total cross section can be obtained via relations (54) and (55) by simply replacing $x_{\nu m}^{a}(t=+\infty)$ by $x_{\nu m}^{d}(t=+\infty)$ in which ν denotes the final state of the system.

Both the adiabatic and diabatic DQMS approaches have been applied to the study of the laser assisted charge transfer process[64]:

$$He^{++} + H(1s) + \hbar\omega \rightarrow He^{+}(n=2) + H^{+} \tag{65}$$

in a two-state approximation, for the velocity range from 10^5 to 10^7 cm/s and the laser intensity in the range of 0.4 to 4.0 TW/cm^2. Figure 17 shows the total charge transfer cross sections of the reaction (65) as a function of the laser intensity I at λ = 3000 Å for V_0 = 1.0 × 10^7 cm/s and 2 × 10^6 cm/s respectively. Four different results are shown: solid curves, the exact coupled (diabatic) DQMS results; dotted curves, the first-order Magnus (M1) approximation results; long-dash-dotted curves, the first-order perturbation (P1) results; and short-dashed curves, the average cross sections.[64] Notice the marked difference of the conventional P1 and exact results. While the P1 curves are linearly dependent upon the intensity, the exact curves and others show strong nonlinear dependence. Furthermore, the P1 method considerably overestimates the cross sections at high intensity. Figure 18 shows the total charge transfer cross sections of the process (65) as functions of the impact velocity V_0 at λ = 3000 Å for I = 0.4 and 1.0 TW/cm^2 respectively. Curve notations are the same as in Fig. 17. The P1 results exhibit strong linear dependence on the reciprocal of the impact velocity, while the other three closely follow each other and show nonlinear dependence. At the higher velocity end all four calculations are in good agreement. The deviation of the approximation results, i.e., P1, M1, and the average from the exact ones becomes pronounced at the lower velocity end.

C. GVV-DQMS method

For the collision between highly stripped ions A^{+q} with hydrogen atoms in fields, the number of coupled DQMS states can go up very rapidly with the structure complexity of the ions A^{+q}. Shown in Fig. 19 are the five field-free quasi-molecular electronic energies ($3d\sigma$, $3d\pi$, $3p\sigma$, $3p\pi$, and $4f\sigma$) of the $(LiH)^{+3}$ system which are needed to include in the study of the charge transfer process:

$$Li^{3+} + H(1s) + n\hbar\omega \rightarrow Li^{+2}(n=3) + H^+ \quad . \tag{66}$$

The corresponding adiabatic DQMS states are shown in Fig. 20, which exhibits the complexity of the problem, namely a large number of channels are coupled to each other at high laser intensities and the avoided crossing patterns are rather complex.

To tackle this large dimensionality as well as multicurve crossing problems, we have recently developed an alternative approach[68] which incorporates the implementation of the generalized Van Vleck (GVV) nearly degenerate perturbation theory[47] into the DQMS formalism. In essence, the GVV technique allows block partitioning of the infinite dimensional Floquet Hamiltonian into a finite-dimensional model space consisting only those nearly degenerate and strongly coupled DQMS states, and thus reduces greatly the number of effective coupled channels. Further, the GVV-Floquet basis provides stable numerical algorithms for minimizing the large field-induced radial couplings (at those avoided crossing regions) without the need to explicitly construct the transformation between the adiabatic and diabatic DQMS basis. This yields a new set (of much smaller size) of coupled GVV-DQMS equations (neither adiabatic nor diabatic) which are particularly convenient for multichannel calculations. The method has been applied to the process, Eq. (66). Up to 15 coupled channels have been included in the study.[68] Extension to more complex systems is in progress.

V. Conclusion

In this article, I have reviewed several recent developments in Floquet theoretical methods for the study of (a) multiphoton excitation of small polyatomic molecules in monochromatic fields, (b) SU(N) dynamical symmetries, multiphoton excitation and nonlinear coherence in N-level systems driven by polychromatic fields, and (c) multi-charged ion-atom collisions in laser fields. Due to space limitations, several other recent developments, such as L^2-non-Hermitian Floquet methods for multiphoton ionization of atoms[69] and multiphoton dissociation of molecules,[70] Floquet-Liouville super-matrix approach[71] for resonance fluorescence, collisional redistribution of radiation, and wave mixings, etc., are not treated here. It is hoped that the Floquet techniques discussed here and elsewhere will provide practical nonperturbative tools for comprehensive ab initio investigation of complex nonlinear optical processes in the near future.

Acknowledgments

This work was supported in part by the Department of Energy, Division of Chemical Sciences and by the Alfred P. Sloan Foundation. Acknowledgment is also made to the Donors of the Petroleum Research Fund, administered by the American Chemical Society, for partial support of this work. The author acknowledges the receipt of a Visiting Fellowship and the hospitality of the Joint Institute for Laboratory Astrophysics, where this manuscript was completed. The work reported here could not have been accomplished without the excellent collaborations of several coworkers, particularly Drs. Tak-San Ho, Cecil Laughlin, and James V. Tietz.

APPENDIX

Analytical Expressions for Multiphoton Transition Probabilities

and Generalized Bloch-Siegert Shifts in Two- and Three-Level

Systems Driven by Bichromatic Fields

A. Notations

Unperturbed system energies: E_α

Bichromatic fields: $\vec{\varepsilon}_i \cos(\omega_i t + \phi_i)$, $i = 1,2$

Coupling strengths: $V_{\alpha\beta}^{(i)} = -\frac{1}{2} \langle \alpha | \vec{\mu} \cdot \vec{\varepsilon}_i | \beta \rangle \exp(i\phi_i)$, $i = 1,2$

$P_{\alpha\to\beta}(t)$: time-dependent transition probability

$\bar{P}_{\alpha\to\beta}$: long time-averaged transition probability

δ_n: generalized Bloch-Siegert shift

B. Two-level system ($E_\alpha < E_\beta$, $\omega_0 = E_\beta - E_\alpha$)

(a) One-photon resonance $\omega_0 \simeq \omega_1$

$$P_{\alpha\to\beta}(t) = \frac{4b^2}{(\omega_0 - \omega_1 + \delta_0) + 4|b|^2} \sin^2 q_0 t \qquad (A1)$$

where

$$\delta_0 = \frac{|b|^2}{\omega_1} + 2|R|^2 |b|^2 \left(\frac{1}{\omega_1+\omega_2} + \frac{1}{\omega_1-\omega_2}\right)$$

$$4q_0^2 = (\omega_0 - \omega_1 + \delta_0)^2 + 4|b|^2 \quad .$$

$$b = V_{\alpha\beta}^{(1)} \exp(-i\phi_1)$$

and

$$R = (V_{\alpha\beta}^{(2)}/V_{\alpha\beta}^{(1)}) \exp[i(\phi_1-\phi_2)] \quad .$$

(b) $(2n+1)$-photon resonance $\omega_0 \simeq (n+1)\omega_1 - n\omega_2$

$$P_{\alpha \to \beta}(t) = \frac{(\frac{1}{2} u_n)^2}{[\omega_0 - (n+1)\omega_1 + n\omega_2 + \delta_n]^2 + (\frac{1}{2} u_n)^2} \sin^2 q_n t \quad , \quad (A2)$$

where

$$\delta_n(\text{shift}) = 2b^2 (\frac{1}{\omega_0 + \omega_1} + \theta(n) \frac{1}{\omega_0 - \omega_1}) + 2R^2 b^2 (\frac{1}{\omega_0 + \omega_2} + \frac{1}{\omega_0 - \omega_2}) \quad ,$$

$$u_n(\text{width}) \propto |R^n b^{2n+1}| \quad ,$$

$$u_1(\text{3-photon}) = 4 \left| Rb^3 (\frac{1}{(\omega_0 - \omega_1)^2} + \frac{2}{(\omega_0 + \omega_2)(\omega_0 - \omega_1)}) \right| \quad ,$$

$$u_2(\text{5-photon}) = 4 \left| R^2 b^5 [\frac{4}{(\omega_0 - \omega_1)^4} + \frac{2}{(\omega_0 - \omega_1)^3} (\frac{1}{\omega_1} + \frac{4}{\omega_0 + \omega_2}) \right.$$

$$\left. + \frac{2}{(\omega_0 + \omega_2)(\omega_0 - \omega_1)^2} (\frac{1}{\omega_1} + \frac{2}{\omega_0 + \omega_2}) - \frac{1}{2} \frac{1}{\omega_1 \omega_2 (\omega_0 + \omega_2)(\omega_0 - \omega_1)}] \right| \quad ,$$

$$4q_n^2 = [\omega_0 - (n+1)\omega_1 + n\omega_2 + \delta_n]^2 + (\frac{1}{2} u_n)^2 \quad ,$$

and

$$\theta(n) = \begin{cases} 0 & \text{if} \quad n = 0 \\ 1 & \text{otherwise} \end{cases} \quad .$$

The expressions (A1) and (A2) are correct to the lowest nonvanishing order only. High-order expressions for $P_{\alpha \to \beta}(t)$ and $\bar{P}_{\alpha \to \beta}$ can be found in Ref. 33.

C. Three-Level System

The three-level system considered here is of the cascade type (Fig. 9) ($E_1 < E_2 < E_3$) with levels 1 and 3 having the same parity but opposite to that of level 2. We shall present the results for the case that $(2n_1 + 1)$ photons of the first field (ω_1) and $(2n_2 + 1)$ photons of the second field (ω_2) are absorbed or emitted, where n_1 and n_2 are arbitrary integers, i.e. $|1\rangle \xrightarrow{(2n_1 + 1)\omega_1} |2\rangle \xrightarrow{(2n_2 + 1)\omega_2} |3\rangle$ is the dominant process.

The physical parameters are the (generalized) Rabi frequencies

$$\alpha_{n_1} = \frac{(-1)^{n_1} \cdot (b_{12}^{(1)})^{2n_1+1} \cdot e^{-i(2n_1+1)\phi_1}}{2^{2n_1} \cdot (n_1!)^2 \cdot \omega_1^{2n_1}} \quad ,$$

$$\beta_{n_2} = \frac{(-1)^{n_2} \cdot (b_{23}^{(2)})^{2n_2+1} \cdot e^{-i(2n_2+1)\phi_2}}{2^{2n_2} \cdot (n_2!)^2 \cdot \omega_2^{2n_2}} \quad ,$$

where

$$b_{\alpha\beta}^{(i)} = -\frac{1}{2} \langle \alpha | \vec{\mu} \cdot \vec{\epsilon}_i | \beta \rangle \quad , \quad i = 1,2 \quad ,$$

and the detunings

$$\Delta_1 = (E_2 - E_1) - (2n_1+1)\omega_1 \quad ,$$

and

$$\Delta_2 = (E_3 - E_2) - (2n_2+1)\omega_2 \quad .$$

At the exact "shifted" multiphoton resonance position, one gets

$$\bar{p}_{1\to2}^{(res)} = \frac{2|\alpha_{n_1}|^2}{(\Delta_1 - \delta_1 + \delta_2)^2 + 4\Omega^2} \quad ,$$

$$\bar{p}_{1\to3}^{(res)} = 2|\alpha_{n_1}|^2 \, |\beta_{n_2}|^2 \cdot \{\frac{1}{\Omega^4} - \frac{1}{\Omega^2[(\Delta_1 - \delta_1 + \delta_2)^2 + 4\Omega^2]}\}$$

where

$$\Omega^2 = |\alpha_{n_1}|^2 + |\beta_{n_2}|^2 \quad ,$$

and the various ac-Stark shifts are given by

$$\delta_1 = -|b_{12}^{(2)}|^2 \cdot \{\frac{1}{(2n_1+2)\omega_1} + \theta(n_1) \cdot \frac{1}{2n_1\omega_1}\}$$

$$- |b_{12}^{(2)}|^2 \cdot \{\frac{1}{(2n_1+1)\omega_1 + \omega_2} + \frac{1}{(2n_1+1)\omega_1 - \omega_2}\} \quad ,$$

$$\delta_3 = |b_{23}^{(1)}|^2 \cdot \{\frac{1}{\omega_1 + (2n_2+1)\omega_2} + \frac{1}{-\omega_1 + (2n_2+1)\omega_2}\}$$

$$+ |b_{23}^{(2)}|^2 \cdot \{\frac{1}{(2n_2+2)\omega_2} + \theta(n_2) \cdot \frac{1}{2n_2\omega_2}\} \quad ,$$

$$\delta_2 = -(\delta_1 + \delta_3) \quad .$$

The corresponding expressions in the RWA limit at the RWA (unshifted) <u>two</u>-photon resonance condition (i.e., $\Delta_1 + \Delta_2 = 0$) are

$$\bar{P}_{1\rightarrow2}^{(res)}(RWA) = \frac{2|b_{12}^{(1)}|^2}{(\Delta_1^2 + 4\Omega'^2)} \quad ,$$

$$\bar{P}_{1\rightarrow3}^{(res)}(RWA) = 2|b_{12}^{(1)}|^2 |b_{23}^{(2)}|^2 (\frac{1}{\Omega'^4} - \frac{1}{\Omega'^2(\Delta_1^2 + 4\Omega'^2)}) \quad ,$$

where

$$\Omega'^2 = |b_{12}^{(1)}|^2 + |b_{23}^{(2)}|^2 \quad .$$

More general expressions for multiphoton non-resonant cases can be found in Ref. 34.

REFERENCES

1. Recent reviews can be found, for example, in Nonlinear Behavior of Molecules, Atoms, and Ions in Electric, Magnetic or Electromagnetic Fields, edited by L. Neel (Elsevier, Amsterdam, 1979); Coherent Nonlinear Optics, edited by M. S. Feld and V. S. Letokhov (Springer, New York, 1980); Photoselective Chemistry, edited by J. Jortner, R. D. Levine and S. A. Rice, Adv. Chem. Phys., Vol. 47 (Wiley, New York, 1981); P. A. Schulz, Aa. S. Suadbo, D. J. Krajnovich, H. S. Kwok, Y. R. Shen and Y. T. Lee, Ann. Rev. Phys. Chem. 30, 311 (1979); V. S. Letokhov, Nonlinear Laser Chemistry (Springer, 1983); Multiphoton Ionization of Atoms, edited by S. L. Chin and P. Lambropoulos (Academic, New York, 1984); Advances in Multiphoton Processes and Spectroscopy, edited by S. H. Lin, Vol. 1 (World Scientific, 1984, 1985).

2. N. Bloembergen and E. Yablonovitch, Phys. Today 31, 23 (1978); A. H. Zewail, V. S. Letokhov, R. N. Zare, R. R. Bernstein, Y. T. Lee and Y. R. Shen, in a special issue on Laser Chemistry, Phys. Today 33, 25-59 (1980).

3. J. H. Shirley, Phys. Rev. 138, B979 (1965).

4. D. T. Pegg, J. Phys. B 6, 246 (1973); S. Stenholm, J. Phys. B 6, 1650 (1973).

5. For a review of Floquet developments before 1976, see D. R. Dion and J. O. Hirschfelder, Adv. Chem. Phys. 35, 265 (1976).

6. For a review of most recent developments in Floquet theory (1976-1984), see, S. I. Chu, Adv. At. Mol. Phys. 21, 0000 (1985).

7. G. Floquet, Ann. de l'Ecole Norm. Sup. 12, 47 (1983).

8. J. H. Poincaré, Les Methodes Nouvelles de la Mechanique Celeste, Vols. I, II, IV, Paris (1892, 1893, 1899).

9. S. H. Autler and C. H. Townes, Phys. Rev. 100, 703 (1955).

10. V. I. Ritus, Sov. Phys.-JETP 51, 1544 (1966).

11. Ya. B. Zel'dovich, Sov. Phys.-JETP 51, 1492 (1966).

12. C. Cohen-Tannoudji and S. Haroche, J. Phys. (Paris) 30, 153 (1969).

13. F. H. M. Faisal, Nuovo Cim. B 33, 775 (1976).

14. S. I. Chu, J. V. Tietz and K. K. Datta, J. Chem. Phys. 77, 2968 (1982). Figure 1 of this paper gives the Floquet matrix structure for the combined (ac+dc) problem.

15. J. V. Tietz and S. I. Chu, in Spectral Line Shapes, Volume 2, edited by K. Burnett (Walter de Gruyter, New York, 1983), pp. 863-877.

16. R. Duperrx and H. van den Bergh, J. Chem. Phys. 73, 585 (1980).

17. P. Gozel and H. van den Bergh,, J. Chem. Phys. 74, 1724 (1981).

18. R. B. Bernstein, Editor, Atom-Molecule Collision Theory (Plenum, New York, 1979).

19. P. A. Braun and A. N. Petelin, Sov. Phys. JETP $\underline{39}$, 775 (1974); V. P. Makarov and M. V. Federov, Sov. Phys. JETP $\underline{43}$, 615 (1976).

20. S. I. Chu, T. S. Ho and J. V. Tietz, Chem. Phys. Lett. $\underline{99}$, 422 (1983); T. S. Ho and S. I. Chu, J. Chem. Phys. $\underline{79}$, 4708 (1983).

21. J. V. Tietz and S. I. Chu, Chem. Phys. Lett. $\underline{101}$, 446 (1983).

22. S. E. Bailkowski and W. A. Guillory, Chem. Phys. Lett. $\underline{60}$, 429 (1979); D. Proch and H. Schröder, Chem. Phys. Lett. $\underline{61}$, 426 (1979).

23. G. L. Wolk, R. E. Weston and G. W. Flynn, J. Chem. Phys. $\underline{73}$, 1649 (1980); T. B. Simpson and N. Bloembergen, Optics Comm. $\underline{37}$, 256 (1981).

24. P. H. Winston, <u>Artificial Intelligence</u> (Addison-Wesley, Reading, MA, 1979).

25. T. B. Simpson and N. Bloembergen, Chem. Phys. Lett. $\underline{100}$, 325 (1983).

26. J. V. Tietz and S. I. Chu, to be published.

27. M. Quack and E. Sutcliffe, Chem. Phys. Lett. $\underline{105}$, 147 (1984).

28. K. F. Milfeld and R. E. Wyatt, to be published.

29. (a) H. W. Galbraith, D. Martin and J. I. Steinfeld, Phys. Rev. A $\underline{26}$, 1528 (1982). (b) P. L. de Vries, C. Chang, T. F. George, B. Laskowski and J. R. Stallcop, Phys. Rev. A $\underline{22}$, 545 (1980). (c) R. V. Ambartzumian and V. S. Letokhov, in <u>Chemical and Biochemical Applications of Lasers</u>, Vol. 3, edited by C. B. Moore (Academic, New York, 1977), pp. 167-314. (d) Y. Zur, M. H. Levitt and S. Vega, J. Chem. Phys. $\underline{78}$, 5293 (1983).

30. N. B. Delone, V. A. Kovarskii, A. V. Masalov and N. F. Perel'man, Sov. Phys. Usp. $\underline{23}$, 472 (1980).

31. T. S. Ho, S. I. Chu and J. V. Tietz, Chem. Phys. Lett. $\underline{96}$, 464 (1983).

32. S. I. Chu and T. S. Ho, Israel. J. Chem. $\underline{24}$, 237 (1984).

33. T. S. Ho and S. I. Chu, J. Phys. B $\underline{17}$, 2101 (1984).

34. K. Wang, T. S. Ho, and S. I. Chu, J. Phys B (in press).

35. H. Salwen, Phys. Rev. $\underline{99}$, 1274 (1955).

36. For most recent works in this field, see <u>Coherence and Quantum Optics</u>, Vol. V, edited by L. Mandel and E. Wolf (Plenum, New York, 1984).

37. Y. Zur and S. Vega, J. Chem. Phys. $\underline{79}$, 548 (1983), and references therein.

38. R. G. Brewer and E. L. Hahn, Phys. Rev. A $\underline{11}$, 1641 (1975); D. Grischkowsky, M. M. T. Loy and P. F. Liao, Phys. Rev. A $\underline{12}$, 2514 (1975).

39. G. Alzetta, A. Gozzini, L. Moi and G. Orriols, Nuovo Cim. B 36, 5 (1976); G. Alzetta, L. Moi and G. Orriols, Nuovo Cim. B 52, 209 (1979); J. E. Thomas, P. R. Hemmer, S. Ezekel, C. C. Leiby, Jr., R. H. Picard and C. R. Willis, Phys. Rev. Lett. 48, 867 (1982); H. R. Gray, R. M. Whitley and C. R. Stroud, Jr., Opt. Lett. 3, 218 (1978); M. Kaivola, N. Bjerre and O. Poulsen, Opt. Comm. 49, 418 (1984).

40. R. M. Whitley and C. R. Stroud, Jr., Phys. Rev. A 14, 1498 (1976); B. Sobolewska, Opt. Comm. 19, 185 (1976); C. Cohen-Tannoudji and S. Reynaud, J. Phys. B 10, 365 (1977).

41. T. W. Mossberg and S. R. Hartmann, Phys. Rev. A 23, 1271 (1981), and references therein.

42. R. P. Feynman, F. L. Vernon and R. W. Hellwarth, J. Appl. Phys. 28, 49 (1957).

43. F. T. Hioe and J. H. Eberly, Phys. Rev. Lett. 47, 838 (1981).

44. J. Elgin, Phys. Lett. 80A, 140 (1980); F. T. Hioe and J. H. Eberly, Phys. Rev. A 25, 2168 (1982).

45. M. Gell-Man and Y. Néeman, The Eight-Fold Way (Benjamin, New York, 1964).

46. T. S. Ho and S. I. Chu, Phys. Rev. A 31, 659 (1985).

47. B. Kirtman, J. Chem. Phys. 49, 3890 (1968); P. R. Certain and J. O. Hirschfelder, J. Chem. Phys. 52, 5977 (1970).

48. T. S. Ho and S. I. Chu, Phys. Rev. A 32, 377 (1985).

49. B. J. Dalton and P. L. Knight, Opt. Comm. 42, 411 (1982); B. J. Dalton and P. L. Knight, J. Phys. B 15, 3997 (1982); S. Swain, J. Phys. B. 15, 3405 (1982); T. A. B. Kennedy and S. Swain, J. Phys. B 17, L389 (1984).

50. M. Kaivola, N. Bjerre, and O. Poulsen, Opt. Comm. 49, 418 (1984).

51. R. P. Hackel and S. Ezekiel, Phys. Rev. Lett. 42, 1736 (1979); C. Delsart and J. C. Keller, J. Physique 39, 350 (1978); R. E. Tench, B. W. Peuse, P. R. Hemmer, J. E. Thomas, S. Ezekiel, C. C. Leiby, Jr., R. H. Picard and C. R. Willis, J. Physique Coll. 42, C8-45 (1981); J. E. Thomas, S. Ezekiel, C. C. Leiby, Jr., R. H. Picard and C. R. Willis, Opt. Lett. 6, 298 (1981).

52. See, for example, R. C. Isler and E. C. Crume, Phys. Rev. Lett. 41, 1296 (1978).

53. (a) R. H. Dixon, J. F. Seely, and R. C. Elton, Phys. Rev. Lett. 40, 122 (1978); (b) A. V. Vinogrudov and I. I. Sobelman, Sov. Phys. JETP 36, 115 (1973); (c) W. H. Louisell, M. O. Scully, and W. B. McKnight, Phys. Rev. A 11, 989 (1975).

54. S. I. Yakovlenko, Sov. J. Quant. Electron. 8, 151 (1978).

55. W. R. Green, M. D. Wright, J. F. Young, and S. E. Harris, Phys. Rev. Lett. 43, 120 (1979).

56. J. F. Seely and R. C. Elton, Naval Research Laboratory, Memorandum Report 4317 (1980).

57. R. Z. Vitlina, A. V. Chaplik, and M. V. Entin, Sov. Phys. JETP $\underline{40}$, 829 (1975).

58. D. A. Copeland and C. L. Tang, J. Chem. Phys. $\underline{65}$, 3161 (1976); $\underline{66}$, 5126 (1977).

59. L. I. Gudzenko and S. V. Yakovlenko, Sov. Phys. Tech. Phys. $\underline{20}$, 150 (1975).

60. E. A. Andreev and A. S. Prostner, Sov. Phys. JETP $\underline{49}$, 998 (1979).

61. J. F. Seely, J. Chem. Phys. $\underline{75}$, 3321 (1981).

62. (a) L. F. Errea, L. Méndez, and A. Riera, J. Chem. Phys. $\underline{79}$, 4221 (1983); (b) L. F. Errea, L. Méndez, and A. Riera, Chem. Phys. Lett. $\underline{104}$, 401 (1984).

63. Y. P. Hsu, M. Kimura, and R. E. Olson, Phys. Rev. A $\underline{31}$, 576 (1985).

64. T. S. Ho, S.-I. Chu and C. Laughlin, J. Chem. Phys. $\underline{81}$, 788 (1984).

65. S.-I. Chu, Chem. Phys. Lett. $\underline{70}$, 205 (1980).

66. (a) T. F. George, J. Phys. Chem. $\underline{86}$, 10 (1982) and references therein; (b) N. M. Kroll and K. M. Watson, Phys. Rev. A $\underline{13}$, 1018 (1976); (c) A. M. Lau, ibid. $\underline{13}$, 139 (1976).

67. (a) F. T. Smith, Phys. Rev. $\underline{179}$, 111 (1969); (b) T. G. Heil and A. Dalgarno, J. Phys. B $\underline{12}$, L557 (1979).

68. T. S. Ho, C. Laughlin, and S. I. Chu, Phys. Rev. A $\underline{32}$, 122 (1985).

69. S.-I. Chu and W. P. Reinhardt, Phys. Rev. Lett. $\underline{39}$, 1195 (1977); S.-I. Chu, Chem. Phys. Lett. $\underline{54}$, 367 (1978); A. Maquet, S.-I. Chu and W. P. Reinhardt, Phys. Rev. A $\underline{27}$, 2946 (1983).

70. S.-I. Chu, J. Chem. Phys. $\underline{75}$, 2215 (1981); S.-I. Chu, C. Laughlin and K. K. Datta, Chem. Phys. Lett. $\underline{98}$, 476 (1983).

71. T. S. Ho and S.-I. Chu, Phys. Rev. Lett., submitted.

Figure Captions

Fig. 1. Structure of the exact Floquet Hamiltonian $\hat{H}_F$ in the Floquet state basis $\{|vj,n>\}$. The Hamiltonian is composed of the diagonal Floquet blocks of type A, and the off-diagonal blocks of type B. $E_{vj}^{(0)}$ are the unperturbed vibrational-rotational energies, and $b_{vj,v'j'}$ are electric dipole coupling matrix elements.

Fig. 2. Quasienergy plots (a), and time-average MPE transition probabilities $\bar{P}_{00\rightarrow vj}$ (b), for the HF molecule subjected to both the laser (E_{ac} = 1.0 TW/cm^2) and the dc electric fields (E_{dc} = 10^{-4} a.u.) simultaneously. Dot-dash lines indicate one-photon peaks, dashed lines two-photon peaks, and solid lines three-photon peaks. Nonlinear effects such as power broadening, dynamical Stark shift, Autler-Townes multiplet splitting, hole burning, and S-hump behaviors, etc., are observed and can be correlated with the quasienergy diagram.

Fig. 3. Structure of the orientation (θ)-dependent vibrational-field Floquet Hamiltonian $\hat{H}_F^{(v)}(\theta)$ in the Floquet state basis $\{|v,n>\}$. $E_v^{(0)}$ are the unperturbed vibrational energies, and $a_{vv'}(\theta)$ are electric dipole coupling matrix elements.

Fig. 4. The approximate band structure (indexed by P) of the quasivibrational energy (OVE) spectrum is shown in the topmost diagram. Only the P = 0 band (i.e., the central G block) needs to be considered in the total quasienergy calculations. Shown here is also the structure of the G matrix in the OVE and rotational basis.

Fig. 5. Time-averaged MPE transition probabilities for the diatomic molecule $^{12}C^{16}O$ initially prepared at states $|v=0,j=3,m_j>$, with (a) m_j = 0, (b) m_j = 1, (c) m_j = 2, and (d) m_j = 3, respectively, at 50 GW/cm^2. — - — - —, one-photon spectra; ·····, two-photon spectra; and ———, three-photon spectra.

Fig. 6. (a) Frequency-dependent total quasi-energies ε_β denoted by β = vj in the figure). (b) and (c) are time-averaged MPA transition probabilities $P_{00\rightarrow vj}$ for CO at 50 GW/cm^2, (a) and (b) are the results obtained by the nonadiabatic approximation whereas (c) is that of exact Floquet calculation. Note the excellent agreement of (b) and (c). The nonadiabatic calculation is about an order of magnitude faster than the exact Floquet calculation in this specific case.

Fig. 7. Dependence of the average number of photons absorbed ($\bar{n}$) by SO$_2$ on the excitation laser intensity. The different traces correspond to the indicated excitation frequencies (cm^{-1}) of the laser. Typical IR MPD of SO$_2$ requires about forty 9.3 μm photons. The results shown in this figure indicate that MPD of SO$_2$ cannot be achieved for I < 20 GW/cm^2.

Fig. 8. Floquet Hamiltonian for two radiation-field problems, constructed in a symmetric pattern. ω_1 and ω_2 are the two radiation frequencies and $V_{\alpha\beta}^{(i)}$ (i = 1,2) are the electric dipole coupling matrix elements for ith field. Note that the diagonal block A possesses an identical Floquet structure as that of one laser problem. Figure 8 can be generalized to the N-field problem.

Fig. 9. Two types of cascade three-level systems have nonvanishing dipole matrix elements between levels 1 and 2, and 2 and 3. (a) Type I has large disparity between the two transition frequencies, i.e., $(E_2-E_1) \gg (E_3-E_2)$, and (b) type II has small disparity between the two transition frequencies, i.e., $(E_2-E_1) \approx (E_3-E_2)$, but $(E_2-E_1) \neq (E_3-E_2)$.

Fig. 10.   Time-dependent transition probabilities (a) $P_{11}(t)$, (b) $P_{12}(t)$, and (c) $P_{13}(t)$ for the cascade three-level system at $(E_2-E_1) = 4000.0$, $(E_3-E_2) = 50.0$, $\Delta_1 = -\Delta_2 = 5.0$, and Rabi frequencies $\Omega_1 = \Omega_1' = \Omega_2 = \Omega_2' = 2.0$ (arbitrary units). The time periods characterizing the population oscillations are defined in terms of reciprocals of the differences between the quasienergies λ_{100} and $\lambda_{2,-1,0}$, and λ_{100} and $\lambda_{3,-1,-1}$, i.e., the one-photon flopping period $T_{12} = 2\pi/|\lambda_{100}-\lambda_{2,-1,0}|$ and the two-photon flopping period $T_{13} = 2\pi/|\lambda_{100}-\lambda_{3,-1,-1}|$. In the RWA calculations, dotted curves, $T_{12}^{(RWA)} = 0.8322$ and $T_{13}^{(RWA)} = 4.9283$, while in the GVV calculations, solid curves, $T_{12} = 0.8362$ and $T_{13} = 4.8866$. Time development of the first four two-photon flopping periods is shown in this case.

Fig. 11.   Components of the subvector $\vec{A}(t)$, i.e., (a) $A_1(t)$, (b) $A_2(t)$, and (c) $A_3(t)$, as functions of time t. Parameters same as shown in Fig. 10.

Fig. 12.   Components of the subvector $\vec{B}(t)$, i.e., (a) $B_1(t)$, (b) $B_2(t)$, and (c) $B_3(t)$, as functions of time t. Parameters same as shown in Fig. 10.

Fig. 13.   Projection of the trajectory of the subvector $\vec{A}(t)$ on the A_1-A_2 plane for the same duration as shown in Fig. 11. (a) GVV results and (b) RWA results.

Fig. 14.   Projection of the trajectory of the subvector $\vec{B}(t)$ on the B_1-B_4 plane for the same case and time duration as shown in Fig. 12. (a) GVV results and (b) RWA results.

Fig. 15.   Projection of the trajectory of the subvector $\vec{B}(t)$ on the B_2-B_3 plane for the same case and time duration as shown in Fig. 12. (a) GVV results and (b) RWA results.

Fig. 16.   The length squares of the subvectors (a) $K_1 = |\vec{A}(t)|^2$, (b) $K_2 = |\vec{B}(t)|^2$, and (c) $K_3 = |\vec{C}(t)|^2$, and of the total coherence vector, $K = |\vec{S}(t)|^2 = \frac{4}{3}$, as functions of time t for the same case as described in Fig. 10. Solid curves, GVV results and dotted curves, RWA results.

Fig. 17. The total charge transfer cross sections of the process (65) as functions of the laser intensity I, in TW/cm^2, at (a) $\lambda = 3000$ Å, $V_0 = 1.0 \times 10^7$ cm/s, and (b) $\lambda = 3000$ Å, $V_0 = 2 \times 10^6$ cm/s. For curve notations see text.

Fig. 18. The total charge transfer cross sections of the process (65) as functions of the impact velocity V_0, in cm/s, at (a) $I = 0.4$ TW/cm^2, $\lambda = 3000$ Å, and (b) $I = 1.0$ TW/cm^2, $\lambda = 3000$ Å. For curve notations see text.

Fig. 19. The field-free quasimolecular electronic energies of the $3d\sigma$, $3d\pi$, $3p\sigma$, $3p\pi$, and $4f\sigma$ states of the $(LiH)^{+3}$ system.

Fig. 20. The adiabatic DQMS energies $\varepsilon^a_{\alpha m}(R)$ of the $(LiH)^{+3}$ + field system at the impact parameter $\rho = 0.1$ a.u., the field intensity $I = 10$ TW/cm^2, and the field wavelength $\lambda = 2000$ Å. Indices appearing on the right-hand side denote the Floquet photon-blocks to which the levels bracketed belong.

218

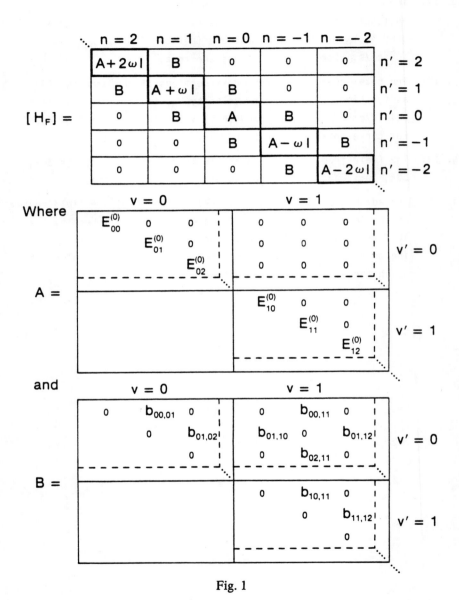

Fig. 1

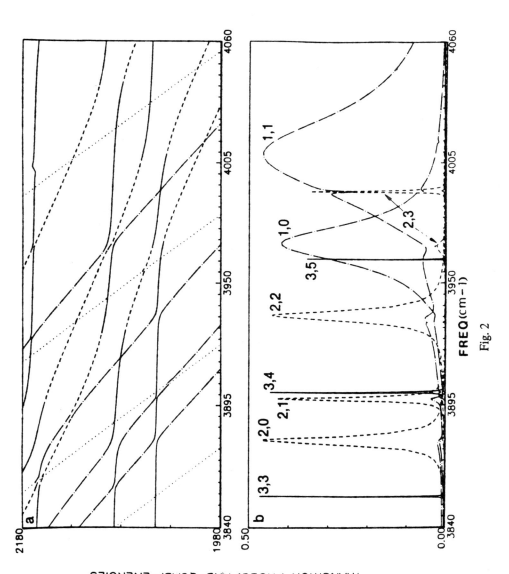

TRANSITION PROBS. AND QUASI-ENERGIES

Fig. 2

	$n = 2$	$n = 1$	$n = 0$	$n = -1$	$n = -2$	
	$A + 2\omega I$	B	0	0	0	$n' = 2$
	B	$A + \omega I$	B	0	0	$n' = 1$
$[H_F^{(v)}(\theta)] =$	0	B	A	B	0	$n' = 0$
	0	0	B	$A - \omega I$	B	$n' = -1$
	0	0	0	B	$A - 2\omega I$	$n' = -2$

Where

	$v = 0$	$v = 1$	$v = 2$	
	$E_0^{(0)}$	0	0	$v' = 1$
$[A] =$	0	$E_1^{(0)}$	0	$v' = 2$
	0	0	$E_2^{(0)}$	$v' = 3$

and

	$v = 0$	$v = 1$	$v = 2$	
	$a_{00}(\theta)$	$a_{01}(\theta)$	$a_{02}(\theta)$	$v' = 1$
$[B(\theta)] =$		$a_{11}(\theta)$	$a_{12}(\theta)$	$v' = 2$
			$a_{22}(\theta)$	$v' = 3$

Fig. 3

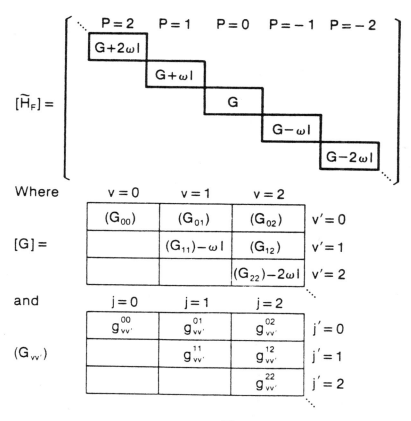

Fig. 4

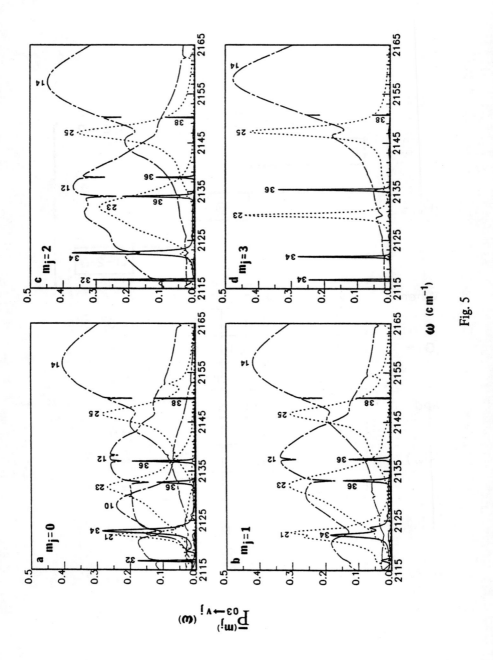

Fig. 5

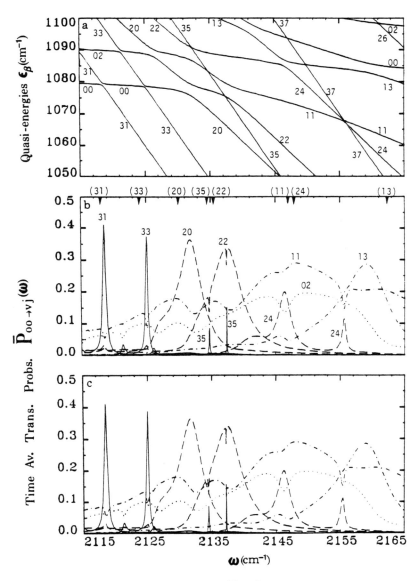

Fig. 6

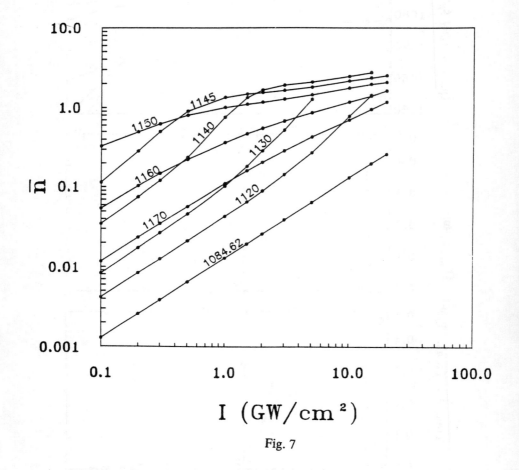

Fig. 7

$$\left[H_F\right]=
\begin{array}{|c|c|c|c|c|}
\hline
A + 2\omega_2 I & B & 0 & 0 & 0 \\
\hline
B & A + \omega_2 I & B & 0 & 0 \\
\hline
0 & B & A & B & 0 \\
\hline
0 & 0 & B & A - \omega_2 I & B \\
\hline
0 & 0 & 0 & B & A - 2\omega_2 I \\
\hline
\end{array}$$

WHERE

$$A =
\begin{array}{|c|c|c|c|c|}
\hline
C + 2\omega_1 I & X & 0 & 0 & 0 \\
\hline
X & C + \omega_1 I & X & 0 & 0 \\
\hline
0 & X & C & X & 0 \\
\hline
0 & 0 & X & C - \omega_1 I & X \\
\hline
0 & 0 & 0 & X & C - 2\omega_1 I \\
\hline
\end{array}$$

$$C =
\begin{pmatrix}
E_\alpha & 0 & 0 \\
 & E_\beta & 0 \\
 & & E_\gamma
\end{pmatrix}
\qquad
X =
\begin{pmatrix}
0 & V_{\alpha\beta}^{(1)} & V_{\alpha\gamma}^{(1)} \\
 & 0 & V_{\beta\gamma}^{(1)} \\
 & & 0
\end{pmatrix}$$

$$B =
\left(
\begin{array}{c|c}
\begin{matrix} 0 & V_{\alpha\beta}^{(2)} & V_{\alpha\gamma}^{(2)} \\ & 0 & V_{\beta\gamma}^{(2)} \\ & & 0 \end{matrix} & O \\
\hline
O & \begin{matrix} 0 & V_{\alpha\beta}^{(2)} & V_{\alpha\gamma}^{(2)} \\ & 0 & V_{\beta\gamma}^{(2)} \\ & & 0 \end{matrix}
\end{array}
\right)$$

Fig. 8

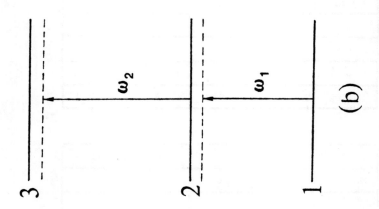

(a)

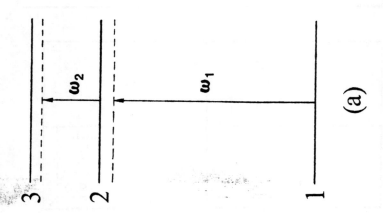

(b)

Fig. 9

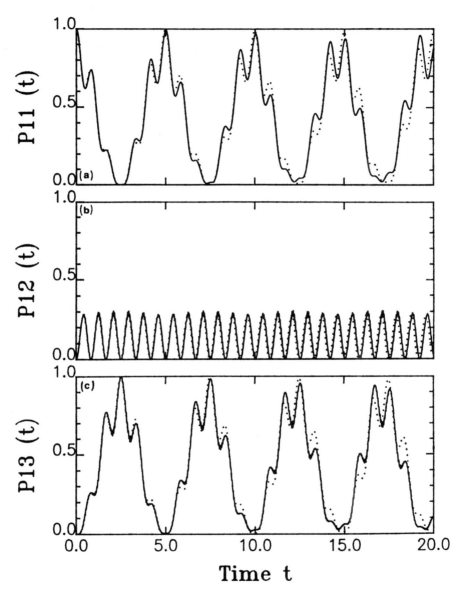

Fig. 10

228

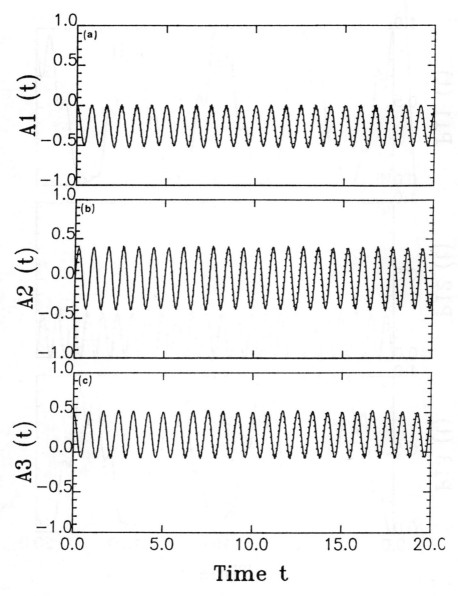

Fig. 11

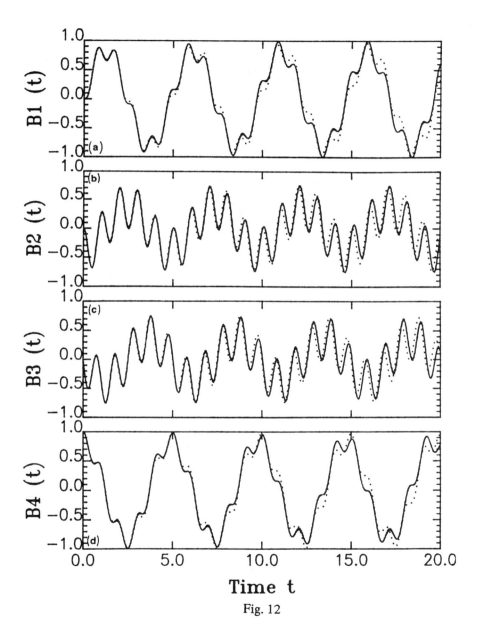

Fig. 12

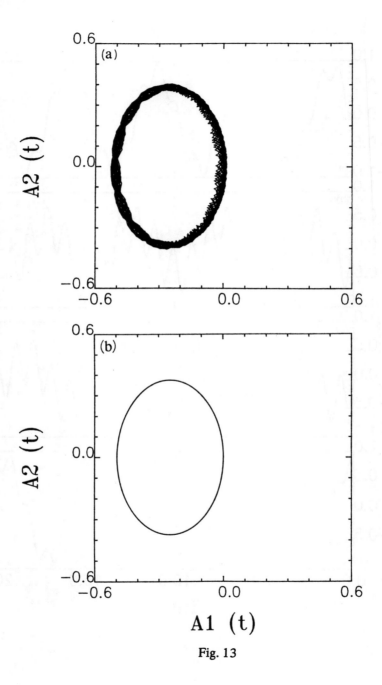

Fig. 13

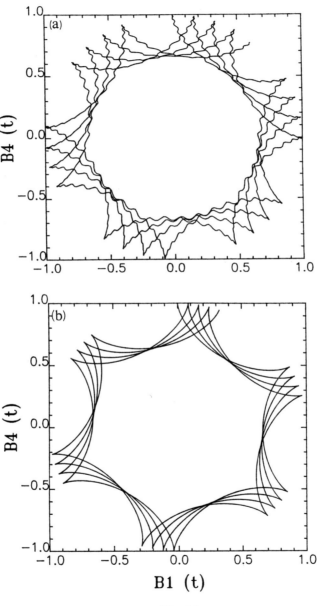

Fig. 14

232

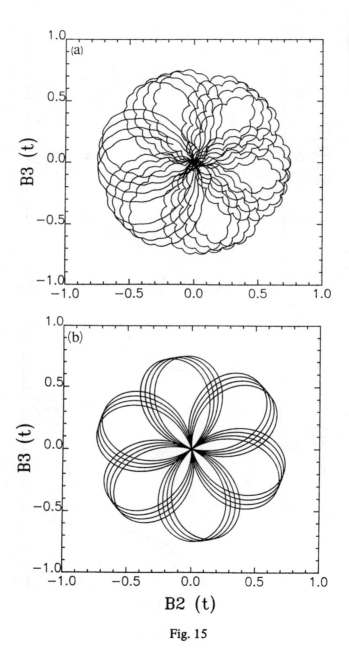

Fig. 15

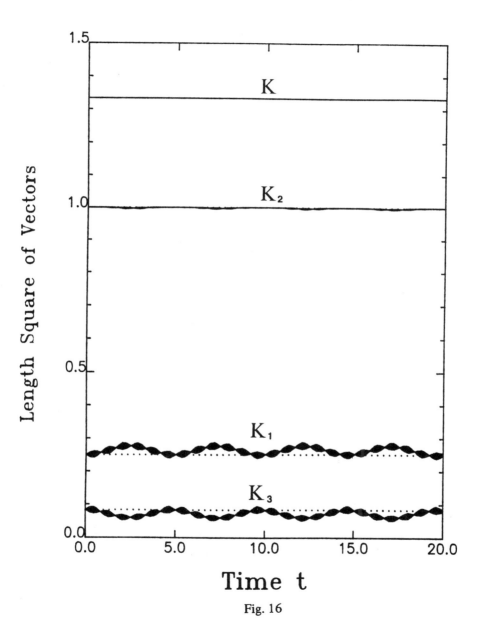

Fig. 16

234

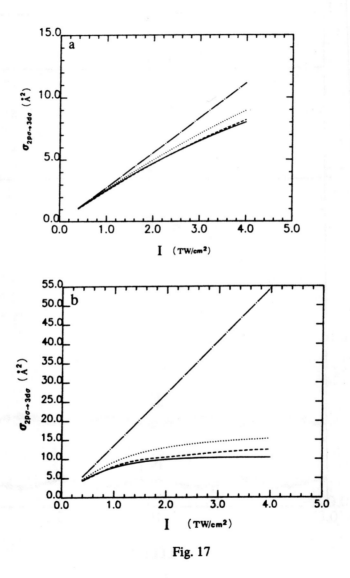

Fig. 17

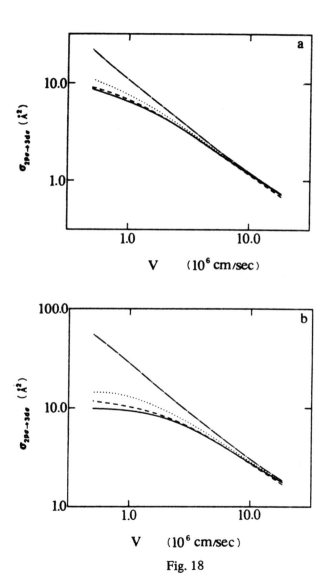

Fig. 18

236

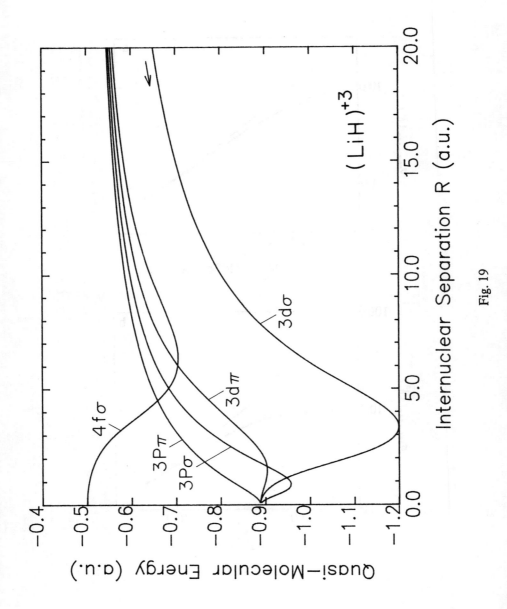

Fig. 19

$(LiH)^{+3}$

Internuclear Separation R (a.u.)

Quasi-Molecular Energy (a.u.)

$4f\sigma$

$3P\pi$

$3P\sigma$

$3d\pi$

$3d\sigma$

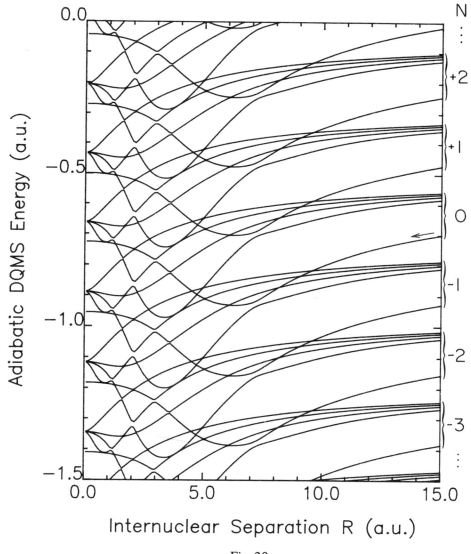

Fig. 20

4. TIME-RESOLVED RESONANCE RAMAN SPECTROSCOPY

Walter Hub, Siegfried Schneider and Friedrich Dörr

Institut für Physikalische und Theoretische Chemie
Technische Universität München
Lichtenbergstrabe 4, D-8046 Garching, GERMANY

Contents

1. Introduction

Time-Resolved Resonance Raman (TR^3) spectroscopy is a special variety of time-resolved vibrational spectroscopy. With this method the vibrational spectra of short-lived molecules can be measured at a selected time during their lifetime. Radicals, radical ions, isomers or other intermediates in chemical reactions as well as molecules in electronically excited states, with lifetimes from picoseconds to milliseconds, may be observed by this method[1-7]. The frequencies and relative intensities of the vibrational bands can be used to obtain detailed structural information about the unstable compounds. In combination with the kinetic information contained in the band intensities, this method is well suited to elucidate the mechanism of unknown chemical reaction pathways. In addition it is possible to get insight into the excited electronic states of the transient species through the measurement of the RR excitation profile.

Other time-resolved vibrational spectroscopic techniques use IR absorption or coherent anti-Stokes Raman scattering (CARS). This review is intended to focus on TR^3 spectroscopy and give a comprehensive overview on the experimental techniques which are presently applied, and on the molecular species

which have been studied so far. Selected typical examples
will be discussed briefly in combination with the experimental
techniques along the line of increasing experimental expense.
For further information a list of references is included,
which is ordered according to the chemical structure of the
compounds studied by this method. The literature survey is
intended to be complete up to june 1985. Finally our own
work on TR^3 spectroscopy will be described.

2. Theory of the resonance Raman effect

In this chapter, a brief introduction into the modern theoretical treatment of the RR effect and the relation between
the electronic absorption spectrum and the RR excitation
profile is given.

If the wavelength of the laser light, which excites the
Raman spectrum of a molecule, coincides with one of its electronic absorption bands, then the intensity of the scattered
light may be intensified by several orders of magnitude as
compared to the non-resonant case. This phenomenon is called
the resonance Raman (RR) effect. It is essential for TR^3
spectroscopy for several respects:
(i) it allows to measure the vibrational spectra of species
at very low concentration, such as reaction intermediates;

(ii) it may allow to observe a RR band of a dissolved compound which happens to have the same frequency as a non-resonant solvent band;

(iii) it selectively enhances the vibrational Raman bands of that compound in a mixture, to which the exciting laser frequency is tuned to resonance. This makes it possible to selectively study low-concentrated transient species in chemical reaction mixtures or, e.g., the chromophore in a biological chromoprotein in the presence of the "uncoloured" macromolecule.

The Raman excitation profile /8/ can provide selective information on the electronic absorption spectrum of the compound. Friedman and Hochstrasser /9/ pointed out that, due to interference effects, deviations from the normal absorption profile can occur. These depend on linewidths and frequencies of the Raman lines; they usually will be below 50 cm^{-1}, for which reason it is generally assumed that the excitation profiles exhibit maxima near the 0-0 and 0-1 transitions of the absorption spectrum. Therfore, the excitation profiles may be used to correlate TR^3 data with those from transient absorption measurements.

For liquid solutions with an isotropic distribution of solute molecules, the total Raman scattered light intensity is given by Placzek's semiclassical theory as /10,11/:

$$I_{i,f} = \frac{2^7 * p^5}{3^2 * c^3} * I_o * \nu^4 * \sum_{\rho,\sigma} |(\alpha_{\rho,\sigma})_{i,f}|^2 \tag{1}$$

where I_o is the intensity of the incident light, $\underline{n}$ is the frequency of the scattered light, $\alpha_{\rho,\sigma}$ is the ρ,σ-th-component of the polarizability tensor, and i,f are the initial and final states, respectively. The modern treatment of inter-action of light with matter is based on the concept of in-duced polarization, which is expanded in a power series of the applied electric field. The sth order electrical suscepti-bility at frequency ω_k can be expressed as the corresponding Fourier component of the induced polarization.

$$P^{(s)}(\omega_k) = \chi^{(s)}(\omega_k = \omega_1 + \omega_2 + \ldots + \omega_s) * E(\omega_1) * E(\omega_2) \ldots E(\omega_s)$$

It is evident that the sth order electrical susceptibility χ is an (s+1)[th] rank tensor. The elements of the tensor are in general complex quantities; the number of independent elements is determined by the point symmetry of the inter-acting material.

As Lee and Albrecht pointed out /12,16/, the different
types of spectroscopy can be classified according to the
order of $\chi^{(s)}$, which is appropriate for their description.
In "active" spectroscopies, the radiation field produces
a well defined change of spectroscopic states of the assembly
of molecules and a net amount of work passes between radiation
field and matter. In "passive" spectroscopies, matter acts
only in a catalytic way to change the state of the radiation
field. "Active" spectroscopies depend on the imaginary part
of the complex susceptibility; they are restricted to the
odd order components of $\chi(s)$; "passive" spectroscopies are
described by both the real and imaginary part of the susceptibility.

In the absence of an resonant intermediate state the long
known expressions for non-resonant Raman scattering /11/
can be used. If, however, an intermediate resonance is ap-
proached, scattering cross-sections are best determined by
means of a density matrix treatment /12/. This approach considers explicitely the relative time ordering of interactions
of bra and ket vectors and the coherence loss between these
/13/. This treatment also gives an answer to the historical
dispute concerning the difference between resonance fluorescence (hot fluorescence) and resonance Raman scattering (for

a detailed discussion, see e.g. /12/).

The full set of electrical susceptibilities up to the third order has been reported by Bloembergen et al. /14,15/. The third order susceptibility expression consists of 48 terms which are distinguished by the arrangement of the energy denominator (sequence of interactions). Only 3 of them have a (near) resonant energy denominator. They are assumed to be sufficient to describe resonance Raman scattering; their form gave rise to the discussion referred to above. These terms are:

$$D_a^{\rho,\sigma} = Im \frac{|<f|D_r|m>*<m|D_s|g>|^2}{(\omega_{mg}-\omega_1+i*\gamma_{mg})*(\omega_{mf}-\omega_2-i*\gamma_{mf})*(\omega_{fg}-\omega_1+\omega_2+i*\gamma_{fg})} \qquad (3a)$$

$$D_{b,c}^{\rho,\sigma} = Im \frac{|<f|D_r|m>*<m|D_s|g>|^2}{(\omega_{mg}-\omega_1\mp i*\gamma_{mg})*(\omega_{mf}-\omega_2-i*\gamma_{mf})*(i*\gamma_{mm})} \qquad \begin{matrix}(3b:-)\\(3c:+)\end{matrix}$$

with

$h\omega_{mg}/2*\pi$ = transition energy between states m and g

γ_{mg} = dephasing constant between states m and g

γ_{mm} = linewidth (inverse lifetime) of state m

According to Bloembergen et al./14,/ the three terms can be rearranged to yield:

$$\chi^{(3)}(\omega_2) \approx \mid \sum_{\rho,\sigma,m} < f|D_\rho|m > * < m|D_\sigma|g > \mid^2 *$$

$$* \frac{\gamma_{fg}*(1+\text{Re } K) - (\omega_{fg}-\omega_1+\omega_2)*\text{Im } K}{[(\omega_{mg}-\omega_1)^2 + \gamma_{mg}^2]*[(\omega_{fg}-\omega_1+\omega_2)^2 + \gamma_{fg}^2]} \tag{4}$$

where

$$K = \frac{i*(\gamma_{mf}-\gamma_{mg}-\gamma_{fg}) + (1-2*\gamma_{mg}/\gamma_{mm})*(\omega_{fg}-\omega_1+\omega_2+i*\gamma_{fg})}{(\omega_{mf} - \omega_2 - i*\gamma_{mf})} \tag{5}$$

Equation (4) states that the product of the transition dipole moments taken between the intermediate electronically excited state m and the initial and final vibronic states i and f, resp., governs the Franck-Condon details of the Raman lines, independently of the actual form of the complex line shape function.

If one neglects first the corrections induced by the factor K, then the second term in (4) clearly shows the resonance effects. The value of $\chi^{(3)}$ becomes resonantly enhanced when

(i) $\omega_1 - \omega_2 = \omega_{fg}$ = vibrational quantum and

(ii) $\omega_1 = \omega_{mg}$ = (electronic) excitation energy to intermediate state m . Inclusion of the factor K leads to an unsymmetric line shape; the effect can, however, be more readily seen by inspection of equs. (3a) and (3b). If ω_1 is tuned away from the electronic transition ω_{mg}, the term D_a shows a resonance for $\omega_2 = \omega_1 - \omega_{fg}$, whilst the second term $D_{b,c}$ shows the resonance at $\omega_2 = \omega_{mf} = \omega_1 - \omega_{fg}$. Since, however, $\gamma_{mf} \gg \gamma_{fg}$, only the D_a term gives rise to a sharp and strongly enhanced resonance Raman band, whilst the terms $D_{a,b}$ comprise only a broad peak of low maximum intensity ("hot luminescence") (for more details see e.g. /17/).

The effect caused by the two types of resonances can of course be seen only if the vibrational levels are widely spaced. In the case of large molecules in condensed phase, the density of vibronic states will be high; to calculate $\chi^{(3)}$ and thereby the Raman scattering cross section, the contributions of the various nearly resonant states must be summed up. This imposes, in addition, the problem of a proper description of the excited states. We shall not go into this problem, but refer the reader to the review by Champion and Albrecht /17/.

For the evaluation of the Raman intensities, i.e. the

calculation of the matrix elements of the dipole moment opera-
tor $<i|D_\rho|m>$ etc., we can rely on the treatment given by
A.C. Albrecht in 1961 /17/ (in that paper, Albrecht has used
the expressions derived for α by Behringer and Brandmüller
/11/, where a damping correction was made by the addition
of $i*\gamma$ in the energy denominator of the Kramers-Heisenberg
dispersion equation).

Albrecht's approach is based on the Herzberg-Teller expansion for vibronic transition dipolmoments. The vibronic
states $\psi(r,q)$ are approximated as products of an electronic
wave function $\theta(r,Q)$ and a vibrational state function $\Lambda(Q)$;
we introduce the following short notation:

$$m \equiv \psi_m(r,Q) = \theta_M(r,Q)*\Lambda_m(Q) = |M>|m_M> \tag{6a}$$

$$f \equiv \psi_f(r,Q) = \theta_G(r,Q)*\Lambda_f(Q) = |G>|f_G> \tag{6b}$$

$$i \equiv \psi_i(r,Q) = \theta_G(r,Q)*\Lambda_i(Q) = |G>|i_G> \tag{6c}$$

For the sake of simplicity, we assume that the vibrational
wavefunctions $\Lambda(Q)$ can be approximated by a product of harmonic oscillator functions, each characterized by the normal
coordinate Q_a and the vibrational quantum number v_a.
Furthermore, the states i, f and m are assumed to differ only

with respect to the vibrational quantum number of one normal coordinate.

Adiabatic Born-Oppenheimer excited states S (M and others) are expressed in a power series of crude Born-Oppenheimer states, S°, taken at the ground state equilibrium coordinates. This procedure results in a power series expansion of the "adiabatic" matrix elements in terms of matrix elements calculated via the crude Born-Oppenheimer wavefunctions. To first order in nuclear displacement, one gets:

$$\langle f|D|m\rangle = \langle f_G| \langle G|D|M\rangle |m_M\rangle \; = \tag{7}$$

$$= \langle f_G|m_M\rangle * \langle G°|D| M°\rangle +$$

$$\sum_{s,a} [\langle G°|D|S°\rangle * \langle f_G|Q_a|m_M\rangle *$$

$$* (\langle M°|dH/dQ_a|S°\rangle / (E°_S - E°_M)]$$

The terms making up the product of dipole moment matrix elements are usually grouped according to the power of the vibronic interaction. This separation allows for an easy discussion of the differences in intensity of totally symmetric and non-symmetric vibrations.

$$\langle i | D_\rho | m \rangle * \langle m | D_\sigma | f \rangle = A_{\rho,\sigma} + B_{\rho,\sigma} \tag{8}$$

with

$$A_{\rho,\sigma} = \langle G^\circ | D_\rho | M^\circ \rangle * \langle M^\circ | D_\sigma | G^\circ \rangle * \langle i_G | m_M \rangle * \langle m_M | f_G \rangle \tag{9}$$

and

$$B_{\rho,\sigma} = \sum_{S,a} [(\langle M^\circ | dH/dQ_a | S^\circ \rangle)/(E^\circ_M - E^\circ_S) *$$

$$* \langle i_G | Q_a | m_M \rangle * \langle m_M | f_G \rangle *$$

$$* (\langle G^\circ | D_\rho | S^\circ \rangle * \langle M^\circ | D_\sigma | G^\circ \rangle +$$

$$\langle G^\circ | D_\sigma | S^\circ \rangle * \langle M^\circ | D_\rho | G^\circ \rangle)] \tag{10}$$

The term A, which is independent of the nature of the nonresonant states S° will contribute to the Raman signal only, if the potential hypersurface in the groundstate G° and in the resonant electronic state M° are different with respect to the normal mode under observation. For totally symmetric normal modes (and only for those), this difference can be realized by a change in equilibrium geometry (distance).

With other words, a totally symmetric mode will produce an
intense Raman band only, if the force constant along this
coordinate is significantly different in the ground and excited electronic state, respectively. For the same reason,
different electronic states will in general provide different
enhancement factors for the various modes. In molecules which
possess at least one symmetry element, the two integrals
$\langle G^\circ|D|M^\circ\rangle$ and $\langle G^\circ|D|S^\circ\rangle$ are not necessarily non-zero at the
same time. (The transition moment vector is directed either
parallel or orthogonal to the symmetry axis (plane)). In
those cases, it will be mainly the diagonal terms of $A_{\rho,\sigma}$
which contribute to the Raman signal. The electronic factor
will be proportional to the square of the transition dipole
moment $|\langle G^\circ|D|M^\circ\rangle|^2$, which also governs the electronic
absorption spectrum; the depolarization ratio reaches a
maximum.

The B-term (equ.10) depends mainly on the vibronic mixing
of the resonant state M° with other excited states S°. Because
of the energy denominator, the contribution is the larger,
the closer in energy the states S° are. In case of molecular
systems with some symmetry, a number of selection rules follow
from inspection of equations (8) - (10):

i) Only those electronically excited states S can contribute to the term B, which carry a non-vanishing transition dipolmoment $\langle G^\circ | D | S^\circ \rangle$ to the ground electronic state.

ii) In order for the vibronic coupling element $\langle M^\circ | dH/dQ_a | D^\circ \rangle$ not to vanish, the product of irreducible representations $G(M^\circ) * G(Q_a) * G(S^\circ)$ must contain the totally symmetric species. For non- degenerate states, $G(S^\circ) = G(M^\circ) * G(Q_a)$ holds. By the symmetry of the electronic wavefunction the direction of the transition dipole moment is fixed with respect to the molecular frame. For a special normal coordinate Q_a and a resonant state M°, the non- vanishing components $B_{\rho,\sigma}$ are those, for which $G(\rho) = G(M^\circ)$ and $G(\sigma) = G(S^\circ) = G(Q_a) * G(M^\circ)$, and vice versa.

The product of overlap integrals $\langle i | Q_a | m \rangle * \langle m | f \rangle$ will be especially large, if the intermediate vibronic state m is one with $v_a = +1$. Due to the presence of the several types of energy denominators, the contribution of each intermediate state will be weighed differently for the Raman active normal coordinates. As a general rule it is observed that the relative intensities vary drastically, if one switches gradually from the (normal) non-resonant to the resonant Raman effect. Further changes occur when, by variation of incident frequency, the photon energy becomes resonant to vibrational levels of different electronic states.

For isotropically oriented molecules, the expression for the depolarization ratio has been given by Placzek /10/ as:

$$\rho = I_o/I_p = (3G'+8G'')/(10G°+4g'+4G'') \qquad (11)$$

where

$$G° = 1/3 * (\alpha_{xx} + \alpha_{yy} + \alpha_{zz})^2$$
$$G' = 1/3 * [(\alpha_{xx} - \alpha_{yy})^2 + (\alpha_{xx} - \alpha_{zz})^2 + (\alpha_{yy} - \alpha_{zz})^2]$$
$$G'' = 1/3 * [(\alpha_{xy} - \alpha_{yx})^2 + (\alpha_{xz} - \alpha_{zx})^2 + (\alpha_{yz} - \alpha_{zy})^2]$$

As mentioned above the A-term (equ.9) contribution should dominate in case of resonance with strongly allowed electronic transitions. Far-off resonance and in case of resonance with weak or electronically forbidden transitions, the B term contribution should govern the total intensity and the depolarization ratio. The latter can, therefore, be used to differentiate between the two contributions or the verify the assumed molecular symmetry /19/ (especially if the molecule under study pocesses a certain symmetry).

From what is said above, it is evident that there is no linear relationship between the absorption spectrum and the (resonance)

Raman excitation profile (REP). For a long time this problem
has been ignored, but with the advent of powerful transform
techniques, even in the case of medium-to-large molecules,
the problem of linking absorption spectra and Raman excitation
profiles has been attacked (see eg. /20-24/). Working with
a series of "standard assumptions", the time correlator technique gives an exact formulation at all temperatures and
for all electron-phonon coupling strengths.

For the purpose of identification of photo-induced transient species, the differences between absorption spectra
and Raman excitation profiles are, however, of minor importance. In order to be able to record a resonance Raman spectrum
of a low concentration intermediate, this intermediate must
have a strongly allowed electronic transition, i.e. the A-term
will dominate. The Franck-Condon factor relevant for absorption,
$|\langle i|m\rangle|^2$, is modified in the Raman excitation profile by
a second overlap integral, $|\langle m|f\rangle|^2$. This may cause small
shifts in the position of the maximum and changes in the
band contour. Despite of this, the band position should be
characteristic enough to give additional evidence for the
assignment of the intermediate in case that several transients
could be proposed from general chemical considerations.

3. Experimental Techniques for Time-Resolved Resonance Raman Spectroscopy

Various experimental arrangements for measuring Raman spectra
in very short times (down to ps) and their combination with
a variety of techniques used to initiate a chemical reaction
at a well defined time have already been reviewed by Atkinson
/2/, Hester /6/ and Delhaye et al. /4,30/, covering the literature up to 1982. In the meantime, experimental progress
has not been very extensive, but the number of research groups
using TR^3 spectroscopy has continued to increase. In the
next section, recent developments will be described together
with basic experimental arrangements and typical results.
Some techniques that have been reviewed already by other
authors, have been included here in order to give a comprehensive review.

3.1 Experiments using a single laser

TR^3 spectra of short-lived photoproducts may be taken using
a single cw laser both the creation and the detection of
the transient species /69,236,237/. In this case the intermediate compound is created by a photoreaction initiated
by the laser light which simultaneously excites the Raman
spectrum of the solution. If the sample is moving through

the laser beam – either in a jet, a flow cell or a rotating cell – kinetic information about the lifetime of the transient species may be obtained. The time resolution of the experiment is determined by the interaction time of the illuminated sample volume with the laser light. This time depends on the collection optics for the scattered light in front of the spectrometer, on the dimension of the cell and on the flow speed. The latter is a free experimental parameter and allows time-dependent measurements. Using flow speeds of 20 m/s in a free jet and a tightly focussed laser beam, the interaction time of the flowing sample with the excitation light may be less than 100 nsec /236/. The scattered light is composed of the Raman or RR spectra of the reactive compound(s), the transient species, the stable products of the reaction and the solvent.

The discrimination between vibrational bands from transient or stable products of the reaction and those from the reactive compound(s) can be made due to the following difference: The intensity of a vibrational band of a transient or stable product of the photoreaction in these experiments depends on the square of the laser intensity whereas there is a linear dependence of the Raman band intensity of the starting compound(s) and the solvent on the laser intensity. Therefore, the contribution of transient or product bands relative to educt and solvent bands in the Raman spectrum will be higher at high photon densities (focussed laser beam) than at lower photon densities (defocussed laser beam). So, if the intensity of a band in the RR spectrum of a solution is growing significantly more than a solvent band in the same spectrum with increasing laser power density, then there is a high probability that this band belongs to a photochemical product of the absorbing species. In conventional RR spectroscopy, photoproduct formation is an undesired perturbation of the spectrum and one has to be careful to avoid a too high photon flux and eventually reduce the laser power for the specific measurement. For example, Asano et al. /83/ measured the absorption of the solution as a function of the laser intensity in order to get a measure of the concentration of the excited states, which are populated under resonance Raman conditions by the incident laser light.

Long-lived intermediates in the photocycle of bacterio-rhodopsin have been studied by Marcus and Lewis /69/ with this method. They concluded that the M_{412} intermediate is isomerized and deprotonated with respect to the starting compound BR_{560}. Intermediates in thermal chemical reactions produced under steady-state conditions using a flow technique, have also been reported /29,31,151/ (see also chapter 3.2.3.).

A different experimental setup uses a single <u>pulsed laser</u>
for photolysis and probe of the transient species /236/.
If the repetition rate is low ($\leq$ 200 Hz with uv laser pumped
dye lasers) a scanning spectrometer would be too slow for
recording the spectra. Shorter accumulation times can be
obtained using the multiplex advantage. The simultaneous
registration of a range of several hundreds of wavenumbers
of a Raman spectrum with each laser pulse can be achieved
by the combination of a spectrograph (instead of a monochro-
mator) and an optical multichannel analyzer using a vidicon
or a diode array instead of a photomultiplier. A typical
experiment is shown in Figure 1 /236/. The time resolution
is given by the width of the laser pulse which normally can
not be varied over a wide range. Therefore, one gets only
information about the reactivity of a substrate within this
laser pulse duration; it is not possible to measure the kinetics of the transient species with this experiment. The flow
speed of the sample needs not to be as high as in the experiment using a single cw laser, but it must be high enough
to replace the photolyzed sample volume by a fresh one before
the next pulse hits the cell. This is extremely important
if irreversible photochemical reactions occur in the solution. The discrimination between the spectra of transient
or stable photoproducts and the starting material is achieved
by the variation of the laser pulse intensity as in the case

described above. Additional and independent (e.g. preparative photochemical) work may be necessary to decide, whether or not stable photoproducts contribute to the measured spectrum. If the pulse energy is too high, nonlinear processes such as multiphoton ionization and dissoziation, stimulated Raman scattering or consecutive photochemical reactions /67/ will destroy the theoretical square dependence of the RR intensity of the transient species on the laser power. This problem is avoided, if one uses a lower laser pulse energy and at the same time increases the repetition rate of the laser pulses in order to reduce the accumulation time of a spectrum. With a cavity- dumped dye laser, synchronously pumped by a mode-locked argon ion laser, in combination with a very fast flow speed of the sample, picosecond TR^3 spectra of bacteriorhodopsin /28/ and hemoglobin /26,27/ photointerme-diates have been measured.

Using ps and ns pulses, Terner et al. /26,27/ observed that spin conversion at the central iron atom takes place in the excited state of carbonmonoxy-hemoglobin during a 30 ps pulse (Figure 2). The newly created state then persists at least for 20 ns. The same method was applied to study the oxy-hemoglobin photoproducts /32/. Raman spectroscopy with single high power laser pulses was used to measure the RR spectra of transient species, such as excited states /34-36/

or changes in the tertiary structure of biomolecules /33/.
The intermediates may have lifetimes as short as femtoseconds
/241/.

Although there have been many valuable contributions to
the characterization of short-lived photochemical intermediates by TR^3 spectroscopy using a single laser, there are
some factors which prevent a broader application:
(i) in order to profit by the resonance enhancement, only
selected photoreactions can be studied, where the educt and
the transient species both absorb in the same wavelength
region;
(ii) the time-resolution is determined by the pulse width
of the laser and may not be varied; with a flowing sample
and a cw laser, it is limited to the microsecond time scale;
(iii) it is not possible to measure RR excitation profiles
(see chapter 2) of the transient species. Changing the laser
wavelength would also result in a change of the concentration
of the intermediate product. In addition, the range of overlapping absorptions of educt and product may not be sufficiently broad to allow the measurement of the full transient
absorption profile.

3.2 Pump-probe experiments

In order to overcome these limitations one has to use a pump-probe arrangement with a second independently triggered source for the creation of the transient species. Several methods have been used so far: photolysis by a second laser system, pulse radiolysis, rapid chemical mixing. In principle, also other techniques for the initiation of dynamical processes in molecules, such as synchron radiation, gamma irradiation, pressure or temperature jump methods, could be combined with a pulsed laser Raman spectrometer.

3.2.1 Photolysis

A schematic diagram of our apparatus for the measurements of nanosecond TR^3 spectra of photochemical transients with two pulsed lasers is shown in Figure 3 /37/. A nitrogen laser pulse (Lambda Physik M 2000; 337 nm, 5 mJ, 5 ns) initiates a photochemical reaction in a sample volume which is slowly flowing through a quartz cell. After an electronically adjustable delay time (30 ns up to 500 ms) a dye laser pulse (excimer laser pumped dye laser: Lambda Physik EMG 500 and FL 2000; typically 5 mJ, 20 ns at 500 nm) excites the Raman spectrum of the solution. The scattered light is collected by a f/1 optics and focussed onto the slit of a double grating spectrograph (Coderg UF76) equipped with a 600 l/mm and a 1800 l/mm grating, respectively. A section of about 600 cm^{-1} of the Raman spectrum imaged at the exit of the spectrograph is intensified by an image intensifier tube (EMI 9912), focussed on a vidicon (PAR 1205), digitized, and stored by the optical multichannel analyzer console (PAR Model 1205A). The photocathode of the image intensifier is externally gated /38/, so that it is active only during a time window, which is synchronized with the probe laser pulse. The tube is nontransmissive especially at that time when the photolysis laser pulse initiates the photoreaction in the solution and eventually creates fluorescence in the same spectral region in which the Raman light is to be recorded. Without gating, high light levels would lead to intense afterglow of the phosphor screens within the image intensifier, thus blurring the very weak TR^3 signal.

The gated intensifier and the vidicon may be replaced by an intensified vidicon or an intensified photodiode array detector which had not been available in 1976 when our setup was designed. Intensified vidicons and photodiode arrays differ in a couple of properties /2/, such as sensitivity, dynamic range, cross talk between adjacent channels, and the linearity and the speed of signal readout. At present, intensified diode array detectors seem to be preferred for one-dimensional detection of spectra in the visible region, whereas the vidicon-type detectors are mainly used in two-dimensional applications and in the infrared region.

Other pump-probe arrangements use nitrogen /40-42/, excimer /43-45/ or Nd-YAG lasers /46-51/ for the photolysis and Nd-YAG lasers /40,46-48,51/ or dye lasers pumped by Nd-YAG lasers /43,49,50/ or excimer lasers /41,42,44,45/ for the Raman probe pulse. Another approach relies on a single pulsed Nd-YAG laser, with the third harmonic of the fundamental being used as the photolysis laser beam and the optically delayed second harmonic as the probe beam /39/. Two cw lasers have been used for measurements under steady-state conditions in a flow-cell /237/ or in a spinning cell at cryogenic temperatures /52/. A broad applicability to various photoreactions is achieved with the use of two independent dye lasers /33,97/.

Recently, Gustafson et al. /58/ published a pump-probe experiment using two independently tunable synchronously pumped and cavity dumped picosecond dye lasers (Figure 4). The output of the first laser was amplified using a cavity dumped argon ion laser and then frequency doubled to yield 10 nJ uv pulses at 0.76 MHz repetition rate. These pulses initiated the photoreaction. The second synchronously pumped cavity dumped dye laser served as the Raman probe (592,7

nm, 10 nJ) and reached the sample after it passed a variable
optical delay line to achieve time-delays from -100 ps to
+70 ps between the pump and the probe pulse. The scattered
light is detected by an intensified reticon array (PAR 1420)
at the output of a single spectrograph (ISA HRA 320). A special technique of accumulating the spectra reduced the baseline drift and allowed to measure a band at 197 cm^{-1} with
very low intensity (13 cps) on a background of much higher
light level (2500 cps) with S/N = 2 /54,57/. The setup was
used to record the Stokes and anti-Stokes RR spectrum of
trans-stilbene in the excited singlet state.

3.2.2 Pulse radiolysis

Electron pulse radiolysis is an alternative method to photolysis for the creation of short-lived radicals, radical ions
or molecules in excited states in the liquid phase. It has
some advantages over photolysis, the most important one being
the possibility to selectively create either the triplet
state or the radical ion of a molecule, independently of
its electronic absorption properties. Therefore, overlapping electronic bands of different substrates in the ground-
state do not lead to problems as may be the case with photolysis /7/. On the other hand, due to the creation of Cerenkov
light while the beam of electrons passes through the cell,

it is not possible to monitor the RR spectrum of a transient
species earlier than a couple of hundred nanoseconds after
the pump pulse.

The first TR^3 spectrum of a short-lived free radical was
recorded with the pump-pulse technique by Pagsberg et al.
in 1976 using a single electron pulse from a Febetron accelerator
to create the radicals and a 600 ns dye laser pulse to excite
the RR spectrum /60/. Meanwhile, electron accelerators working
at higher repetition rates are available. With these it is
possible to accumulate a large number of pump-probe sequences
in order to improve the S/N ratio of the RR spectrum /7/
(Figure 5). Short-lived species with lifetimes above 30 ns
can now be detected /7/.

A great number of transient species produced by pulse
radiolysis in solution have been studied by TR^3 spectroscopy:
(i) triplet states of p-terphenyl /59/, beta-carotene /124,125/,
retinal /113/ and other polyenes /114,115/;
(ii) radicals of the phenoxyl type /86,90,91/;
(iii) radical ions of p-terphenyl /60/, p-benzosemiquinones
/84,87,92,93/, anilin /62/, cytochrome c /112/, dihalides
/61,63/, and radical ion dimers /109,119/.

3.2.3 Rapid mixing

Raman spectroscopic investigations of bimolecular thermo-
chemical reactions in solution can be performed with cw laser
excitation in a conventional scanning Raman spectrometer.
The reactants are rapidly mixed in a chamber and the RR spectrum of the solution is recorded under steady-state conditions
in a capillary at the outlet of the mixing chamber. Long-lived
($\sim$ms) intermediates of chemical reactions such as enzyme-sub-
strate complexes /3/, ion radicals /29,65/, and a reduced
form of cytochrome c /151/ could be observed using this rapid-
flow technique /66/. With a stopped-flow device and multichannel detection /71,72/, or with a rapid scan technique
/71/, the kinetics of product formation of slow chemical
reactions have been investigated.

3.3. Kinetic analysis of TR^3 data

The kinetic analysis of time-dependent RR band intensities
was first described by Atkinson et al. /73/ and later extended
to second order reactions by Hub et al. /37/. In TR^3 experiments, the intensity of a specific RR band of the transient
species is not simply proportional to its concentration at
a given time, due to the fact that the absorption of the
solution is varying with time. The laser pulse intensity

is changed in the solution due to partial absorption by the transient species; the intensities of the non-resonant Raman bands of the solvent reflect the time-dependence of this transient absorption. A linear relationship between the transient RR band intensity and the concentration of the transient species at a given time is achieved by normalizing the RR band intensity to the intensity of a solvent band measured simultaneously under the same conditions /73/. Multichannel detection thus has a great advantage over scanning devices with respect to kinetic analysis of the data. If vibrational bands of ground-state molecules, e.g. solvent or parent molecules, overlap transient bands, a subtraction of the Raman spectrum recorded without the photolysis source (spectrum A) from the spectrum with the transient bands (spectrum B) is necessary in order to be able to accurately calculate the band intensities of the transient species in the difference spectrum $C=B-n*A$. The intensities in spectrum A have to be multiplied by an empirical factor n such that in the difference spectrum C all solvent bands are equally well compensated. This procedure may not be satisfactory for a complete RR spectrum, especially if the electronic absorption of the transient species shows a strong variation over the wavelength range covered by the RR spectrum /73/. In this case, the constant factor n with which the data of spectrum A have to be multiplied should be replaced by the absorption

profile n(λ) of the transient species or by adequate factors
adjusted sequentially for different regions.

4. Current topics

After it had been demonstrated by Wilbrandt and coworkers
in 1975 and 1976 that the RR spectra of a stable radical
(diphenylpicrylhydrazyl /68/) and of a short-lived radical
ion (p-terphenyl anion /60/) can be measured, TR^3 spectroscopy has been applied to study many other short-lived species,
such as molecules in excited states (first triplet spectrum
by Wilbrandt et al. in 1978 /59/), ion radicals, radicals,
isomers and tautomers of organic, inorganic, biological or
biochemical compounds. In Table 1, the transient species
which have been studied until june 1985 are compiled.

The selectivity of the RR probe technique (see chapter 2)
makes TR^3 spectroscopy a powerful method for the study of
biological and biochemical intermediates. By simply tuning
the probe laser wavelength to an electronic absorption band
of a certain transient species it is possible to selectively
detect its RR spectrum without severe perturbations caused
by the presence of other compounds such as proteins, water
or enzymes with different chromophores. At present, equivalent information about the structure of biological inter-

mediates at the reaction site of enzymes etc., as is provided
by TR^3 spectroscopy, is not available from any absorption
technique using uv, visible or IR light. A review on the
application of RR and TR^3 spectroscopy to photosynthesis
has been published recently by Lutz /74/. Therefore we shall
refer to other topics.

4.1. Bacteriorhodopsin and model compounds

Essential structural properties of six intermediates in the
photocycle of bacteriorhodopsin, with lifetimes between picoseconds and milliseconds, were determined by TR^3 spectroscopy.
In Figure 6 are indicated the intermediates and their characteristic structural features, which have been deduced from
transient absorption techniques and from T^3 spectroscopy.
It is generally accepted that the main structural changes
of the chromophor comprise the reversible 13-_trans_ to 13-_cis_
isomerization and a change of the state of protonation of
the Schiff base, which links the chromophor to the surrounding
apoprotein bacterioopsin. Isotopic substitution in the retinal
was a major tool to deduce the structural changes in the
different intermediates by TR^3 spectroscopy /52,75/. Kinetic
RR studies recently revealed an equilibrium between the intermediates L_{550} and M_{412} /76/. In the absence of water intermediates with different electronic absorption and RR spectra

were detected /77,78/ (Figure 7). Modifications of the normal
cycle are also found at high laser intensities /242/. In
this case the illumination leads to interceptions of the '
normal cycle, by which the BR_{570} form is recovered from L
and M via new species L' and M'. These new intermediates
have been studied by transient absorption and by TR^3 spectroscopy.

First RR spectroscopic studies of the photocycle of halorhodopsin (hR) /79/ have been reported /80/. Halorhodopsin
is the photoreceptor of <u>halobacterium halobium</u> for the light--
induced chloride pump across the membrane, in analogy to
bacteriorhodopsin (b), which has the same function for the
proton-pump. The photochemical cycle of hR is not yet established. An intermediate with a deprotonated Schiff base corresponding to bM_{412} is not found in hR. Both in the presence
and in absence of chloride, one and the same short-lived
photoproduct (100 μsec time-scale) was found by TR^3 spectroscopy. Its spectrum is very similar to that of bL_{550}.

In the course of these investigations the vibrational
spectra of the isomers of retinal and of some carotenoid
model compounds became important. The studies have been extended to include the triplet states of carotenoids and diphenylpolyenes of varying chain lengths (see Table 1). Stilbene

has the   shortest chain in this series. It is the molecule
from which the greatest number of unstable photoproducts,
i.e. the excited singlet state, the anion, the dianion and
the cation radical, have already been characterized by TR^3
spectroscopy (see Table 1). The triplet state RR spectrum
however is still unknown; the knowledge of the triplet confor-
mation would be useful for the understanding of the excited
state dynamics of trans-stilbene. Lately, we were able to
record the TR^3 spectra of the planar triplet states of two
trans-4- nitrostilbenes /82/.

4.2. Hemoglobin

The study of hemoglobin photolysis by TR^3 spectroscopy is
another application to photobiology. Progress has been made
in the understanding of different processes following
photolysis: A formerly proposed change in the core-size of
the porphyrin ring is questioned now /5/; instead, an electronically excited high-spin heme is proposed to explain
the spectral changes observed after the excitation of oxy-
hemoglobin with high intensity, 30 ps pulses /32/. In the
excited state the iron is assumed to take a position outside
of the plane of the porphyrin ring, where it is located in
the ground state. A RR band around 1360 cm^{-1} has been found
to be sensitive in its frequency to variations in the π-elec-

tron distribution. This in turn may depend on the oxidation state of the iron or on the surrounding of the porphyrin. Recent studies on this problem were made by Friedman /239/ and by Spiro and Terner /85/. The frequency shifts of this electron-density marker band around 1360 cm^{-1} are inversely correlated to the frequencies of the iron-histidine stretching modes of both the equilibrium species and the photolytically created transients observed on the nanosecond time scale /5/. So it seems as if there is a direct link between the Fe-His stretching mode and the π-electron distribution in the porphyrin ring. The nature of the specific protein-porphyrin interaction which is reflected in this relationship, still is obscure /5/. A model of the influence of the iron-histidine tilting motion on cooperativity and on specific variations of oxygen affinity is presented by Friedman /239/. The relaxation of the tertiary structure in photodissociated hemoglobins from various sources was studied under the influence of pH and temperature /33/ (Figure 8). A model accounting for solution-dependent variations in ligand reactivity within a given quaternary structure of the hemoglobin was developed on the basis of these results. Recently ferryl myoglobin /88/ and horseradish peroxidase /89,94/ also have been studied by TR^3 spectroscopy.

4.3. Excited states

TR^3 spectroscopy has successfully been applied to deduce the structure of a considerable number of molecules in electronically excited states (see Table 1). The singlet and/or the triplet states of polyenes,carotenoids and diphenylpolyenes have already been mentioned before. It can be expected that by a systematic interpretation of their RR spectra in combination with normal mode calculations the dynamics of the isomerization via the triplet route in these compounds may be fully understood in the near future.

A similar collection of data is already available of the triplet states of various polyacene molecules /73,95-97/, like naphthalene /46/, anthracene /42,45,98/, phenanthrene /97/, chrysene /73,98/, acridine /45,99/ and phenazine /99/.

Transition metal bipyridyl complexes $M(bpy)_3^{3+}$ presently are of interest because of their possible use in solar energy storage devices. Their electronically excited states have been characterized by TR^3 spectroscopy; evidence was found for localized MLCT states in these molecules (see Table 1).

4.4. Radicals and radical ions

Electron transfer reactions are amongst the most important reactions in chemistry and biology. Preparative aspects of photoinduced electron-transfer have been reviewed by Mattes and Farid /129/. Understanding of chemical reaction mechanisms relies on the identification of the intermediate products of the reaction, the determination of their structures and their reactivities towards other substrates, and the kinetics of their formation and decay.

So far, only few of the TR^3 spectra of radicals and radical ions listed in Table 1 have been kinetically analyzed. In most cases the characterization of the transient species by its RR spectrum was the only scope of the study, occasionally additional model calculations have been made to reveal a structure. Phenazine radical cations and some derivatives have been studied by Hester at al. /110,111/. Intermediate redox compounds in model systems for solar energy conversion /35,36,64,83,120-123,126,207,217/ and a series of phenoxyl intermediates and semiquinone derivatives also have been published (see Table 1). Other TR^3 spectroscopic studies of transients in electron-transfer reactions have only been done rarely. As an example for the study of both aspects of TR^3 spectroscopy, our own work on the stilbene ion radicals will be discussed in more detail in the next chapter.

In photochemically induced electron-transfer reactions three types of radical ion pairs may be distinguished (Scheme 1) due to their different kinetic or spectroscopic behavior: the contact radical ion pair (A^-D^+) or exciplex $(AD)^*$, the solvent separated radical ion pair $(A^- + D^+)$ and the free radical ions.

<u>Scheme 1</u> General scheme of intermediates in photolytically induced electron-transfer reactions

$$A^* + D \xrightarrow{\quad} {}^1(A^-D^+)^* \xrightarrow{\quad} {}^1(A^- + D^+) \rightleftarrows A^- + D^+$$

$$A + D \quad \longleftarrow \quad {}^3(A^- + D^+)$$

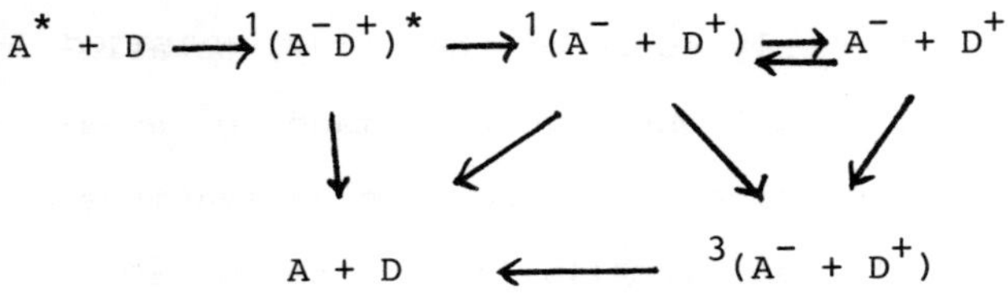

Exciplexes are usually formed only in solvents of low polarity. Some exciplexes are fluorescent at room temperature with radiative lifetimes up to several nanoseconds. The basic ideas about intermolecular electron-transfer mechanisms have been developed by fluorescence quenching and lifetime measure-

ments /128,132,137/. The room temperature absorption or emission
spectra of exciplexes in solution are broad and without vibrational structure. This has prevented TR^3 spectroscopic investigations of exciplexes so far. The time-resolved CARS method
is ideal for this purpose and it has successfully been applied
/145,146/. Solvent-separated radical ion pairs and free solvated radical ions are formed in solvents of high polarity.
As they are non-fluorescent and have lifetimes between 1 ns
and several us in room temperature solution their kinetics
can only be measured by transient absorption, time-resolved
Raman, or CARS spectroscopy. A spectroscopic characterization
of short-lived ion radicals can be achieved by ESR and chemically induced dynamic nuclear polarization (CIDNP). The interaction energy between the radical ions in a solvent-separated
radical ion pair is approximately equal to the solvation
energy of each free radical ion in solvents of high polarity.
Therefore the electronic absorption spectra of a radical
ion in one or the other solvation state will not be significantly different. This could be observed recently by picosecond
transient absorption measurements on the trans-stilbene/fumaro-
nitrile exciplex in polar and non-polar solvents /247/. Hence,
for the identification of transient species in photochemically
induced electron-transfer reactions it is necessary to monitor
their kinetics.

If no other chemical reactions occur, exciplexes and solvent-separated radical ion pairs decay according to a first order law (log c $\sim$ t) whilst the separated radical ions decay according to a second order rate law (1/c $\sim$ t). Typical lifetimes in room temperature solutions are nanoseconds or less for exciplexes and solvent-separated radical ion pairs and microseconds for free radical ions (due to the rather slow diffusion process in liquid solution).

5. Stilbene ion radicals

Our own work in the field of TR^3 spectroscopy focusses on the illumination of unknown photochemical reaction pathways and the determination of the structure of their intermediate products. We will review in this chapter the results of our TR^3 studies of the anion /103,105/ and the cation /37/ radical of trans-stilbene (TS) and its isotopically labelled derivatives /100,244/. Other studies of ion radicals /82,102/ and triplet states /82/ of chemically modified trans-stilbene derivatives will not be included. We have also investigated the electron-transfer reactions of trans-stilbene in non-polar solvent /104,141/. The identification of the stilbene anion radical and the stilbene cation radical and the analysis of their decay kinetics in polar solvents by TR^3 spectroscopy provided the key information about the mecha-

nisms of the electron-transfer reactions. Steady-state
photochemical investigations have been performed additionally
by Lewis and coworkers /103,105/.

5.1 The mechanism of the stilbene-amine photoaddition

At the beginning of our investigation it was known that some
tertiary alkylamines add to trans-stilbene upon irradiation
in polar solutions only and that this addition is stereo-
electronically controlled /130/. Exciplex fluorescence was
found for a few tertiary amines in non-polar solvents /131,132/.
A three-step mechanism was proposed for the addition reaction
by Lewis et al. /131,133/: (i) electron transfer, (ii) proton
transfer, (iii) free radical recombination and disproportio-
nation. One main result of our investigations /103/ is that
the proton transfer takes place in the initially formed exciplex in competition with dissociation of the exciplex.
Another unexpected result is the formation of amine dimer
cation radicals upon quenching of trans-stilbene/amine exciplexes /104,141/ and solvent-separated radical ion pairs
/103/ by amines.

Figure 9a shows the Raman spectrum of a deaerated solution
of trans-stilbene (0.03 M) and ethyldiisopropylamine (0.75 M)
excited at 480 nm with an excimer laser pumped dye laser.

The spectrum consists of bands from the solvent and TS only, amine bands are absent. If the same solution is photolyzed by a nitrogen laser pulse 60 ns prior to the Raman probe pulse then additional bands appear (Figure 9b). The difference spectrum (Figure 9c) is calculated such that the solvent band at 1372 cm^{-1} is completely compensated. The frequencies and relative intensities of the transient spectrum are the same as in the RR spectrum of $TS^{\bar{\cdot}}$ measured by Takahashi and Maeda /134/ and Dosser et al. /135/. The same spectrum and intermediate is found with all amines studied, i.e. ethyl-diisopropylamine (EDA), triethylamine (TEA), 1-azabicyclo[2.2.2]-octane (DABCO) and tetraethyl-1,2-ethanediamine(TEED). It is important to note that $TS^{\bar{\cdot}}$ has been found even with amines (EDA, DABCO, TEED) which do not add to stilbene under these experimental conditions /103/.

The TR^3 spectra of the cation radicals of the amines could not be detected with 480 nm excitation due to their low extinction coefficient. The identification of $TS^{\bar{\cdot}}$ as the transient species is also supported by the coincidence of the RR excitation profile and the main absorption band around 480 nm of $Na^+TS^{\bar{\cdot}}$ /136/. The assignment of the observed frequencies to molecular vibrations will be treated in section 5.3.1.

The decay of the transient species was followed by the dependence of the transient RR band intensities (1577/1553 cm^{-1}) on the delay time between the pump and the probe laser pulses. A second order rate law, $A + A \rightarrow B$, describes the disappearance of TS⁻ bands for all amines measured except tetraethylethylendiamine (TEED) (Figure 10). The lifetime in the latter case was too short to be measured by our apparatus (t ⪈ 40 ns). Both the long lifetime (several hundreds of nanoseconds) and the observed second order rate law show that the transient species is the free solvated TS⁻. Its decay is mainly due to recombination and back electron- transfer with the corresponding amine cation radical.

A deeper insight into the reaction mechanism results from the dependence of the decay kinetics of free TS⁻ and the yields of adduct and cis-stilbene formation on the amine concentration. The yields of both stilbene adduct (determined by conventional photochemical methods), and free TS⁻ (determined as relative TR^3 intensity at t=0 increase with higher solvent polarity and amine substitution (EDA > TEA > TMA). Therefore, both processes seem to be related. However, the yield of adduct formation increases with increasing amine concentration, while the yield of free TS⁻ in TS/TEA solutions does not (Figure 10). The essential conclusion is that free TS⁻ cannot be the direct precursor of the stilbene-amine adduct /103/.

Further investigations of the quantum yield of cis-stilbene
via the triplet route from the singlet radical ion pair showed
that it decreases with increasing TEA concentration similar
to the yield of free TS$^-$. Therefore intersystem crossing
and formation of TS$^-$ may occur from the same intermediate.

Scheme 2 shows the complete mechanism which explains all
experimental results (in polar solutions): The initially
formed exciplex (contact radical ion pair) undergoes hydrogen
transfer to yield 1,2-diphenylethyl-α- aminoalkyl radical
pairs or dissociates into solvent-separated radical ion pairs.
Triplet trans-stilbene and free TS$^-$ is formed from these
solvent-separated radical ion pairs, which also can be quenched
by ground-state amines. All these processes have been proven
experimentally to be dependent upon amine structure and concentration. The decay of the free TS$^-$ is by recombination
with amine cation radical and is independent of amine structure or concentration.

5.2. The reactivity of stilbene cation radicals

According to a calculation of the free enthalpy for electron-
transfer reactions by the Rehm-Weller equation /137/, the
formation of TS$^+$ should be possible by photolyzing solutions
of TS in the presence of various electron acceptors like

fumaronitril (FN) or other organic nitriles. Foote et al.
/138-140/ postulated a mechanism for the photooxygenation
of trans- and cis-stilbene sensitized by 9-cyanoanthracene
and 9,10-dicyanoanthracene (DCA) in polar solvents which
leads via the trans-stilbene cation radical. Experimental
evidence for this interpretation was the observation of a
characteristic transient absorption band around 480 nm by
microsecond flash photolysis /139/.

We have measured TR^3 spectra of solutions of $2 \times 10^{-4}M$
DCA and 0,05 M TS in acetonitrile at room temperature at
delay times between the pump and the probe laser pulses of
50 ns up to 900 ns. A typical transient spectrum is shown
in Figure 11 /37/. The excitation wavelength (485,3 nm) was
close to the maximum of the transient absorption band. The
same transient frequencies are found in photolyzed TS/FN
solutions although a different reaction mechanism is active
there. We assigned the spectra to the free $TS^{\dagger}$ based on the
fact that the transient must be related to the stilbene component
of the reaction mixtures and that the observed lifetimes
was several hundreds of nanoseconds /37/.

A second-order decay of the $TS^{\dagger}$ bands was observed in
all reactions investigated. The decay followed the simple
A+B $\rightarrow$ C+D scheme, where the species A and B have the same

concentration. This is in accordance with a homogeneous recombination of the radical cation and the radical anion. The rate also depended on the oxygen concentration. For TS/FN solutions the decay is greatly enhanced, if the solution is saturated with oxygen instead of nitrogen. Furthermore, benzaldehyde was formed in these solutions with a quantum yield that paralleled the yield of TS^{+}. From both observations it may be concluded that the reaction between TS^{+} and oxygen is much faster than the recombination between TS^{+} and FN^{-}. In the presence of DCA^{-} the decay rate of TS^{+} is almost diffusion controlled in deoxygenated solutions. As a consequence, the influence of dissolved oxygen on the decay rate of TS^{+} in the presence of DCA^{-} is negligeable.

Another interesting detail of the TS/FN reaction mechanism is the formation of fumaronitril anion radical dimers, $(FN)_2^{-}$, for which experimental evidence was found in the dependence of the yield and decay rate of TS^{+} on the concentration of FN. This formation is quite similar to the formation of amine dimer cation radicals in the TS/amine reactions described in the previous chapter. The formation of such ion radical dimers at high concentration of the neutral parent compound seems to be a general reaction competing with the homogeneous recombination of the solvent-separated radical ion pair. Only in a few cases have free organic radical ion dimers been observed RR spectroscopy /142/.

5.3 RR spectra of deuterated trans-stilbene ion radicals

A comparison of the TR^3 spectra of TS, TS^- and TS^+ reveals significant differences, which may be due to different structures arising from different π-electron distribution in these molecules. In addition, the form and the mixing of the normal modes in these molecules may also differ from each other. From MO calculations for TS^- and TS^+ assuming planar geometry, the experimentally observed electronic transition energies and relative transition moments are satisfactorily reproduced /152/. However, the SCF-CI excited state wavefunctions of the resonant transitions (λ=480 nm) are represented by mixtures of singly excited configurations with the main component being due to the equivalent HOMO-LUMO transition in TS. The calculated π-bond orders for TS^- and TS^+ are almost equal in the ground-state and very similar in the excited singlet states. Therefore, one would expect almost identical vibrational frequencies for the planar anion and cation radicals and only small differences in the relative intensities of the RR bands.

Before we can extract structural informations from the RR spectra of TS^- and TS^+ we need to know the assignments of the RR bands to molecular vibrations. To this end we have measured the RR spectra of the ion radicals of several isotopically labelled <u>trans</u>-stilbenes (Figure 12). Isotopic substitution is the only way to solve the problem of assigning an observed band to one of the 36 or 72 possible normal modes (depending on the symmetry of the ion radicals). One normally assumes that the form and the coupling of normal modes remains approximatley unaffected upon increasing the mass of a specific atom in the molecule and that it only changes the frequencies of these vibrational modes which include movement of the isotop. This assumption is good for small changes of the atomic mass like $^{12}C/^{13}C$ exchange, but changes of normal modes upon deuteration may occur. The next two sections describe the RR spectra and assignments of the deuterated stilbene ion radicals.

5.3.1. Trans-Stilbene anion radical

The stilbene anion radicals have been produced by the reaction between the singlet excited <u>trans</u>-stilbenes ($5x10^{-2}$M) and diisopropylethylamine (0,7M) in acetonitrile solution at room temperature. The excitation wavelength was tuned to the maximum of the transient absorption. The RR spectra of some deuterated <u>trans</u>-stilbene anion radicals taken at 70 ns delay, are shown in Figure 13, together with that of the non-deuterated compound /100/.

In the region of the C=C stretching frequencies between 1500 and 1600 cm^{-1} a systematic shift of frequencies occurs. In the TS⁻ and DAA⁻ spectra a strong band at 1577 and 1575 cm^{-1}, resp., is present, whereas in the compounds with fully substituted phenyl hydrogen atoms (D10⁻ and D12⁻) this band is missing. Instead, a new band at 1546 cm^{-1} appears in D12⁻ which could be present in D10⁻, too, but might be covered by the strong 1555 cm^{-1} band. These bands, therefore, may be assigned to the phenylic C=C stretching vibrations (see Table 2). The band at 1553 cm^{-1} in TS⁻ is found at similar position (1555 cm^{-1}) in D10⁻, whereas in DAA⁻ and D12⁻ (both having substituted the hydrogen atoms at the olefinic group in the center of the molecule) new bands appear at 1515 cm^{-1} and 1507 cm^{-1}. These bands are attributable to the olefinic C=C stretching vibrations.

In the region between 1300 - 1000 cm^{-1} there appear two medium intense bands at 1251 cm^{-1} and 1180 cm^{-1} in TS⁻. The higher frequency band is missing in DAA⁻ and D12⁻, but a band at 1243 cm^{-1} in D10⁻ probably corresponds to the same vibration. New bands of comparable intensity appearing only in DAA⁻ and D12⁻ at 987 cm^{-1} and 990 cm^{-1} may be identified as the frequencies belonging to the same vibrational mode. Here the influence of deuteration on the vibrational frequency is strong ($\Delta\tilde{\nu}$ = 260 cm^{-1}); these frequencies are therefore assigned to the CH in plane bending modes of the olefinic part of the molecule. The band at 1180 cm^{-1} in TS$^{\bar{\cdot}}$ is shifted to 1174 cm^{-1} in DAA$^{\bar{\cdot}}$ but more drastically shifted to 851 cm^{-1} in D12$^{\bar{\cdot}}$ and 867 cm^{-1} in D10$^{\bar{\cdot}}$, respectively. We assigne these bands to the phenyl CH in plane bending modes. There is no band found between 1000 and 1300 cm^{-1} in the TS$^{\bar{\cdot}}$ spectrum, which could be assigned to the C-phenyl stretching frequency. Such a band is found at 1193 cm^{-1} in crystalline TS /143/. The assignment of the C-phenyl stretching frequency in the RR spectrum of the <u>trans</u>-stilbene singlet state is still in discussion /58,144/.

The assignments in the frequency range below 1000 cm^{-1} are made by analogy to the observed shifts in the spectra of the parent compounds reported by Meić and Güsten /143/. The one-electron reduction of <u>trans</u>-stilbene obviously does not greatly influence these vibrational frequencies. Bands at 997, 854 and 640 cm^{-1} in TS correspond to bands at 978, 848 and 624 cm^{-1} in TS$^{\bar{\cdot}}$. This assignment is based on the almost equal shifts of these bands upon deuteration of the neutral and the anionic compound.

5.3.2. Trans-Stilbene cation radical

The cation radical spectra of the same deuterated compounds
are shown in Figure 14 /100/. These spectra have been recorded 100 nsec after the photolysis pulse (337 nm) hits the
acetonitrile solution (argon bubbled) of 5×10^{-4} M 9,10-di-
cyanoanthracene and 5×10^{-2} M trans-stilbene. The wavelength
of the probe laser was 480 nm.

The assignment of the bands is given in Table 2. The band
at 1285 cm^{-1} is readily assigned to the vinylic CH defor-
mation mode due its shift to 1018 cm^{-1} in the DAA‡ and D12‡
spectra, and the smaller shift to around 1255 cm^{-1} in the
spectra of those compounds, in which only one of the α-hydro-
gen atoms is replaced by deuterium and finally due to the
fact that the band almost remains unaffected by the other
substitution patterns. Although this assignment is reliable,
there is some uncertainty in the assignment of the 1605 cm^{-1}
band in TS‡. The high intensity of the CH de-formation band
must arise from a mixing of this molecular mode with the
vinylic C=C stretching mode. Therefore, one might expect
the C=C stretching mode to be strong, too. Considering the
possibility that the 1605 cm^{-1} band in TS‡ is the vinyl
stretching mode, we can not find a reasonable explanation
for the double band in the DAA‡ spectrum and the difference
in frequency of the strong bands (1596 and 1567 cm^{-1}) in

the DAA$^+$ and D12$^+$ spectra (the corresponding frequency difference
in the neutral and in the anion radicals are only 11 and
8 cm^{-1}). Also some other characteristic shifts of the vinyl
stretching frequencies upon deuteration are not found in
the anion radical spectra. Therefore it is likely that the
single strong band in the C=C stretching region (except for
DAA$^+$) has to be assigned to the ring C=C stretch. The vinyl
C=C stretch has lost its intensity because of strong coupling
of both C=C stretching modes. This coupling is the stronger
the closer the frequencies are due to the specific deuteration.
In DAA$^+$ the frequency difference is the largest of all spectra
($\Delta \tilde{\nu}$ = 65 cm^{-1}) and therefore both C=C stretching bands are
observed almost unshifted and unaffected with respect to
their intensity. In the spectra of the mono-α-deuterated
cation radicals medium intense bands around 1555 cm^{-1} are
observed which may also be explained by the same argument.
In fact, a similar dependence of the relative intensities
of both C=C stretching bands on the frequency difference
can be observed in the anion radical spectra (Figure 13).
Obviously, this effect is much less pronounced here than
in the cation radical spectra. A detailed discussion of this
point will be given elsewhere /244/.

The ring deformation modes (Table 2) are only weakly
influenced by the one-electron oxidation. In the cation radi-

cal spectra neither the C-phenyl stretch nor the ring CH deformation mode can be identified.

5.4. The structure of stilbene ion radicals

It has been found that both cis- and trans-stilbene can form anion /249,250/, dianion /136/ and cation /249/ radicals. Rapid isomerization of the cis into the trans configuration seems to be common for all these radical ions /136,185,249/. Only in one specific case the cis to trans isomerization is slow /136/. As a conclusion one might expect the ion radicals in a trans configuration which is their thermodynamically most stable form at room temperature. The structures of stilbene ion radicals have not yet been determined. It is still an open question, how much - if ever - the ions are twisted or distorted upon oxidation or reduction of the planar parent molecule.

In ethene a 25° twist for the cation radical was found from the analysis of the vibrational structure in the photo-electron spectrum /147/. MO calculations lead to a variety of predicted twisting angles /148,149/. The carbon-carbon stretching frequency in the cation radical is found at 1250 cm^{-1} /147/, and between 1150 and 1320 cm^{-1} in sterically congested alkenes /150/.

Due to delocalization of the positive or the negative charge, resp., over the whole stilbene skeleton in the cation or anion radical, the C=C stretching frequencies of the vinyl and the phenyl groups are found at lower values than in the spectrum of neutral _trans_-stilbene. The shift of the vinyl C=C stretching frequency, therefore, is particularly smaller in _trans_-stilbene (-89 cm^{-1} and -80 cm^{-1} resp.) than in ethene (-406 cm^{-1}) /244/.

For planar stilbene ion radicals one expects a higher bond order of the C-phenyl bond and lowered bond orders for the vinyl bond and some bonds in the phenyl rings /152/. The bond order may be lowered by a twisting or a rotation of the phenyl group at this bond relativ to the central C=C bond. Unfortunately neither in the anion nor in the cation radical spectra a vibrational band assignable to the C-phenyl stretching vibration is found. A normal mode analysis of _trans_-stilbene radical ions is currently in progress /248/ using the frequencies and assignments of this work /100,244/. It will probably reveal the structures of the ion radicals. A comparison of the RR spectra of the radical ions with those of the excited singlet state of _trans_-stilbene /49,54,57,144,153/, the chemically modified anion and cation radicals of _trans_-stilbene /82,102/, and the triplet states of <u>trans</u>-stilbene derivatives /82/ can demonstrate structural modifications of a molecular skeleton under the influence of changing π-electron densities due to substitution, electron-transfer and electronic excitation.

6. Concluding remarks

TR^3 spectroscopy of unstable compounds is currently a rapidly developing field in Raman spectroscopy. Due to the progress in laser technology and optical detection systems, the wavelength range for the excitation of a TR^3 spectrum will be shifted out into the UV following the current expansion of RR spectroscopy into this wavelength region /154-156/. It then will be possible to study photo-induced structural changes in the protein pocket around biological chromophores by monitoring the changes in the RR spectra of specific amino acids. More experimental improvements to suppress fluorescence in ps TR^3 measurements may be found /246/. With laser pulses shorter than 1 ps, spectral information is lost due to their increased linewidth. Kinetic data may, however, still be evaluated. As alternative methods, time-resolved CARS and ir spectroscopies should be mentioned. They are of special advantage for the study of highly fluorescent transients. Compared with TR^3 spectra, CARS spectra are harder to analyze. The application of ir spectroscopy is limited by solvent and educt absorption and by its lower sensitivity.

Acknowledgments: The authors wish to thank F.D. Lewis, North-western University, Evanston, for a fruitful collaboration and all other co-workers which have contributed to the results presented here. We are indebted to professor Z. Meić, University of Zagreb, Yugoslavia, and Professor H. Güsten, Kern-forschungszentrum Karlsruhe, for their kind gift of samples of deuterated stilbenes. We also want to thank V. Schlegel for careful typing of the mansucript. Our work was supported by the Deutsche Forschungsgemeinschaft and a NATO travel grant, which is greatfully acknowledged. We finally wish to beg the indulgence of those colleagues whose work has been discussed too briefly or altogether neglected.

7. References

1 G. H. Atkinson
   In "Advances in Laser Spectroscopy", B. A.Garetz, J. R. Lombard, Eds.,
   Heyden, London 1982, pp. 155-175

2 G. H. Atkinson
   In "Advances in IR and Raman spectroscopy", R.J.H. Clark, R. E. Hester,
   Eds., Heyden, London 1982, Vol. 9, pp. 1-62

3 P. R. Carey and A. C. Storer
   Ann. Rev. Biophys. Bioeng. 13 (1984) 25-49

4 M. Delhaye, M. Bridoux and F. Wallart
   J. Molec. Struct. 79 (1982) 51-66

5 M. Friedman, D. L. Rousseau and M. R. Ondrias
   Ann. Rev. Phys. Chem. 33 (1982) 471-491

6 R. E. Hester
   Spex Speaker 27 (1982)

7 G. N. R. Tripathi
   ACS symposium series on "Image in Spectroscopy", Y. Talmi, Ed., 1983,
   171-182

8 L. Rimai, R. G. Kilponen and D. Gill
   J. Am. Chem. Soc. 92 (1970) 3824

9 J. Friedman and R. M. Hochstrasser
   Chem. Phys. Lett. 32 (1975) 414

10 G. Placzek,
   In: "Handbuch der Radiologie", G. Marx, Ed., Akademische Verlagsgesell-
   schaft Leipzig, 1934, Vol. I2, pp. 209

11 J. Behringer and J. Brandmüller
   Z. Elektrochem. 60 (1956) 643

12 D. Lee and A. C. Albrecht J. Chem. Phys. 78 (1983) 3382;
   Y. Fujimara and S.H. Lin, J. Chem. Phys., 70, 247 (1979); 71, 3733 (1979).

13 D. Lee and A. C. Albrecht
   In :"Advances in IR and Raman Spectroscopy", Vol. 12, Heyden, London
   1985; Y. Fujimura and S.H. Lin, J. Chem. Phys., 82 1246 (1985).

14 N. Bloembergen, H. Lotem and R. T. Lynch Jr.
   Ind. J. Pure Appl. Phys. 16 (1978) 151

15 N. Bloembergen
   In: "Laser Spectroscopy IV", H. Walther, K. W. Rothe, Eds., Springer,
   Berlin 1979

16 D. Lee and A. C. Albrecht
   Phys. Rev. A18 (1978) 159

17 P. Champion and A. C. Albrecht
   Ann. Rev. Phys. Chem. 33 (1982) 353

18 A. C. Albrecht
   J. Chem. Phys. 34 (1961) 1476

19 A. Warshel and P. Dauber
   J. Chem. Phys. 66 (1977) 5478

20 V. Hiznyakov and I. Tehver
    Phys. Stat. Sol. 21 (1967) 755

21 D. L. Tonks and J. B. Page
    Chem. Phys. Letters 66 (1979) 449

22 J. B. Page and D. L. Tonks
    J. Chem. Phys. 75 (1981) 5694

23 B. R. Stallard, P. R. Callis, P. M. Champion and A. C. Albrecht
    J. Chem. Phys. 78 (1983) 712

24 D. Lee, B. R. Stallard, P. M. Champion and A. C. Albrecht
    J. Phys. Chem. 88 (1984) 6693-6696

25 G. H. Atkinson, Ed.,
    "Time-Resolved Vibrational Spectroscopy". Academic Press, New York 1983

26 J. Terner, T. G. Spiro, D. F. Voss, C. Paddock and R. B. Miles
    In "Picosecond phenomena III", K. B. Eisenthal et. al., Eds., Springer
    1982, 327-330

27 J. Terner, J. D. Stong, T. G. Spiro, M. Nagumo, M. Nicol and M. A.
    El-Sayed
    Proc. Natl. Acad. Sci. USA 78 (1981) 1313-1317

28 M. A. El-Sayed, C.-L. Hsieh and M. Nicol
    In ref. 25, 251-262

29 E. Ernstbrunner, R. B. Girling, W. E. L. Grossman and R. E. Hester
    J. Chem. Soc. Perkin Trans. II, 1978, 177-184

30 M. Bridoux and M. Delhaye
    In "Advances in IR and raman spectroscopy" Vol. 2, R. J. H. Clark and
    R. E. Hester, Eds., Heyden 1976

31 R. E. Hester, K. P. J. Williams
    J. Chem. Soc. Perkin II 1982, 559-563

32 J. Terner, D. F. Voss, C. Paddock, R. B. Miles and T. G. Spiro
    J. Phys. Chem. 86 (1982) 859-861

33 T. W. Scott and J. M. Friedman
    J. Am. Chem. Soc. 106 (1984) 5677-5687

34 M. R. Ondrias, T. W. Scott, J. M. Friedman and V. W. Macdonald
    Chem. Phys. Letters 112 (1984) 351-355

35 S. McClanahan, T. Hayes and J. Kincaid
    J. Am. Chem. Soc. 105 (1983) 4486-4487

36 Y. C. Chung, N. Leventis, P. J. Wagner and G. E. Leroi
    J. Am. Chem. Soc. 107 (1985) 1416-1417

37 W. Hub, S. Schneider, F. Dörr , J. D. Oxman and F. D. Lewis
    J. Phys. Chem. 88 (1984) 2308-2315

38 F. Dörr   W. Hub, S. Schneider and J. R. Taylor
    EMI Application Note R/S 019 (1977)

39 H. Hamaguchi, H. Okamoto, M. Tasumi, Y. Mukai and Y. Koyama
    Chem. Phys. Letters 107 (1984) 355-359

40 J. Teraoka, H. Hashimoto, S. Matsudaira and Y. Koyama
    Chemistry Letters (1985) 311-314

41 J. N. Moore, G. H. Atkinson, D. Phillips, P.M. Killough and R.E.Hester
    Chem. Phys. Letters 107 (1984) 381-384

42 G. N. R. Tripathi and M. R. Fisher
    Chem. Phys. Letters 104 (1984) 297-302

43 R. Wilbrandt, H.-H. Jensen and F. W. Langkilde
    Chem. Phys. Letters 111 (1984) 123-127

44 R. Wilbrandt, W. E. L. Grossman, P. M. Killough, J. E. Bennett and R.E.
    Hester
    J. Phys. Chem. 88 (1984) 5964-5971

45 P. H. M. Van Zeyl, C. A. G. O. Varma and G. Vroege
    Chem. Phys. Letters 105 (1984) 127-133

46 J. Prasad and P. N. Prasad
    Chem. Phys. Letters 113 (1985) 344-346

47 G. T. Babcock, J. M. Jean, L. N. Johnston, G. Palmer and W. H. Woodruff
    J. Am. Chem. Soc. 106 (1984) 8305-8306

48 T. Kobayashi and S. Koshihara
    Chem. Phys. Letters 104 (1984) 174-178

49 H. Hamaguchi, C. Kato and M. Tasumi
    Chem. Phys. Letters 100 (1983) 3-7

50 G.H. Atkinson, J.B. Pallix, T.B. Friedman, D.A. Gilmore and R. Wilbrandt
    J. Am. Chem. Soc. 103 (1981) 5069-5072

51 S. M. Beck and L. E. Brus
    J. Chem. Phys. 75 (1981) 4934-4940

52 S.O.Smith,A.B.Myers,J.A.Pardoen,C.Winkel,P.P.J.Mulder,J.Lugtenburg,R.A.
    Mathies
    Proc. Natl. Acad. Sci. USA 81 (1984) 2055-2059

53 We are currently using two excimer laser pumped dye lasers (Lambda
    Physik: EMG 500 with FL 2000 and EMG 101 MSC with FL 2002).

54 T. L. Gustafson, D. M. Roberts and D. A. Chernoff
    J. Chem. Phys 79 (1983) 1559-1564

55 A. Laubereau and M. Stockburger, Eds.:
    "Time-Resolved Vibrational Spectroscopy - Emil-Warburg -Symposium",
    Springer, Berlin 1985

56 Organizing Committee for the IXth International Conference on Raman
    Spectroscopy, Ed., "Proceedings of the IXth International Conference on
    Raman Spectroscopy", Tokyo, Japan 1984

57 T. L. Gustafson, D. M. Roberts and D. A. Chernoff
    J. Chem. Phys. 81 (1984) 3438

58 T. L. Gustafson, D. A. Chernoff, J. F. Palmer and D. M. Roberts
    Proceedings of SPIE, Vol. 533, 1985, pp. 78-86

59 R. Wilbrandt, N. H. Jensen, P. Pagsberg, A. H. Sillesen and K. B. Hansen
    Nature 276 (1978) 167-168

60 P. Pagsberg, R. Wilbrandt, K. B. Hansen and K. V. Weisberg
    Chem. Phys. Letters 39 (1976) 538-541

61 R. Wilbrandt, N.-H. Jensen, A. H. Sillesen and K. B. Hansen
    Chem. Phys. Letters 106 (1984) 503-507

62 G. N. R. Tripathi and R. H. Schuler
    Chem. Phys. Letters 110 (1984) 542-545

63 G. N. R. Tripathi, R. H. Schuler and R. W. Fessenden
    Chem. Phys. Letters 113 (1985) 563-568

64 M. Forster and R. E. Hester
    Chem. Phys. Letters 81 (1981) 42-47

65 R. E. Hester and K. P. J. Williams
    J. Chem. Soc. Faraday Trans. 2, 78 (1982) 573-584

66 R. E. Hester
    In "Advances in IR and Raman Spectroscopy", Vol. 4, R. J. H. Clark and
    R. E. Hester, Eds., Heyden, London 1978, pp. 1

67 I. W. Wylie and J. A. Koningstein
    J. Phys. Chem. 88 (1984) 2950-2953

68 R. Wilbrandt, P. Pagsberg, K. B. Hansen and C. V. Weisberg
    Chem. Phys. Letters 36 (1975) 76-78

69 M. A. Marcus and A. Lewis
    Science 195 (1977) 1329-1330

70 E. E. Ernstbrunner, R. B. Girling, W. E. L. Grossman and R. E. Hester
    J. Chem. Soc. Faraday Trans. II 74 (1978) 501-508

71 J. C. Merlin, J.-L. Lorriaux and R. E. Hester
    J. Raman Spectrosc. 11 (1981) 384-389

72 M. Crunelle-Cras and J. C. Merlin
    J. Raman Spectrosc. 6 (1977) 261-263

73 G. H. Atkinson, D. A. Gilmore, L. R. Dosser and J. B. Pallix
    J. Phys. Chem. 86 (1982) 2305-2310

74 M. Lutz
    In "Advances in IR and Raman Spectroscopy", R.J.H. Clark, R. E. Hester,
    Eds., Wiley Heyden, New York 1984, Vol. 11, pp. 211-300

75 R. A. Mathies
    In "Spectroscopy of Biological Molecules", C. Sandorfy, T. Theophanides
    Eds., Reidel Publ., New York 1984, pp. 303-328

76 T. Alshuth and M. Stockburger
    In ref. 55

77 P. Hildebrandt and M. Stockburger
    Biochem. 23 (1984) 5539

78 P. Hildebrandt, A. Hagemeier and M. Stockburger
    In ref. 55

79 W. Stoeckenius and R. A. Bogomolni
    Annu. Rev. Biochem. 52 (1982) 587

80 A. Maeda, Y. Ogurusu, T. Yoshizawa and T. Kitagawa
    submitted 1984

81 J. Terner, T. G. Spiro, M. Nagumo, M. F. Nicol and M. A. El-Sayed
    J. Am. Chem. Soc. 102 (1980) 3238-3239

82 C. Richter and W. Hub
    In ref. 55

83 M. Asano, J. A. Koningstein and D. Nicollin
    Chem. Phys. 73 (1980) 688-696

84 G. N. R. Tripathi
    J. Chem. Phys. 74 (1981) 6044-6049

85 T. G. Spiro and J. Terner
    In ref. 25, 297-306

86 G. N. R. Tripathi and R. H. Schuler
    J. Phys. Chem. 88 (1984) 1706-1710

87 R. H. Schuler, G. N. R. Tripathi, M. F. Prebenda and D. M. Chipman
    J. Phys. Chem. 87 (1983) 5357-5361

88 A. J. Sitter, C. M. Reczek and J. Terner
    Biochem. Biophys. Acta, submitted november 1984

89 J. Terner and D. E. Reed
    Biochem. Biophys. Acta 789 (1984) 80-86

90 G. N. R. Tripathi and R. H. Schuler
    J. Chem. Phys. 81 (1984) 113-121

91 G. N. R. Tripathi and R. H. Schuler
    Chem. Phys. Letters 98 (1983) 594-596

92 G. N. R. Tripathi and R. H. Schuler
    J. Phys. Chem. 87 (1983) 3101-3105

93 G. N. R. Tripathi and R. H. Schuler
    J. Chem. Phys. 76 (1982) 2139-2146

94 J. Terner, A. J. Sitter and C. M. Reczek
    Biochem. Biophys. Acta, IN PRESS 1984

95 G. N. R. Tripathi
    In ref. 56, pp. 328-329

96 G. H. Atkinson and L. R. Dosser
    J. Chem. Phys. 72 (1980) 2195-2197

97 D. A. Gilmore and G. H. Atkinson
    In ref. 25, 161-166

98 S. M. Beck and L. E. Brus
    J. Chem. Phys. 72 (1981) 1031-1033

99 R. J. Kessler, M. R. Fisher and G. N. R. Tripathi
    Chem. Phys. Letters 112 (1984) 577-579

100 C. Richter and W. Hub
    In ref. 55

101 U.Klüter , W. Hub and S. Schneider
    In ref. 55

102 W. Hub
    In ref. 56, 26-29

103 W. Hub, S. Schneider, F. Dörr , J. D. Oxman and F. D. Lewis
    J. Am. Chem. Soc. 106 (1984) 708-715

104 W. Hub, S. Schneider, F. Dörr , J. D. Oxman and F. D. Lewis
    J. Phys. Chem. 87 (1983) 4351-4353

105 W. Hub, S. Schneider, F. Dörr , J.T. Simpson, J.D. Oxman and F.D. Lewis
    J. Am. Chem. Soc. 104 (1982) 2044-2045

106 W. Hub, M. Melzig, S. Schneider and F.Dörr
    Ber. Bunsenges. Phys. Chem. 85 (1981) 505-507

107 F.Dörr , W. Hub and S. Schneider
    J. Molec. Struct. 60 (1980) 233-238

108 W. Hub, S. Schneider and F. Dörr
    Angew. Chem. Int. Ed. Engl. 18 (1979) 323-324

109 R.Wilbrandt, N.-H. Jensen, P.Pagsberg, A.H. Sillesen, K.B. Hansen, R.E.
    Hester
    Chem. Phys. Letters 60 (1979) 315-319

110 R. E. Hester and K. P. J. Williams
    J. Raman Spectrosc. 13 (1982) 91-95

111 R. E. Hester and K. P. J. Williams
    J. Chem. Soc. Perkin Trans. II 1981, 852-859

112 B. Cartling and R. Wilbrandt
    Biochem. Biophys. Acta 637 (1981) 61-68

113 R. Wilbrandt and N.-H. Jensen
    J. Am. Chem. Soc. 103 (1981) 1036-1041

114 R. Wilbrandt and N.-H. Jensen
    Ber. Bunsenges. Phys. Chem. 85 (1981) 508-511

115 F. W. Langkilde, R. Wilbrandt and N.-H. Jensen
    Chem. Phys. Letters 111 (1984) 372-378

116 M. Forster, R. B. Girling and R. E. Hester
    J. Raman Spectrosc. 12 (1982) 36-48

117 S. M. Beck and L. E. Brus
    J. Chem. Phys. 76 (1982) 4700-4704

118 R. Wilbrandt and N.-H. Jensen
    In ref. 25, 273-285

119 R. Wilbrandt, N.-H. Jensen, P.Pagsberg, A.H. Sillesen, K. B. Hansen, R.
    E.Hester
    J. Raman Spectrosc. 11 (1981) 24-26

120   S. M. Angel, M. K. DeArmond, R. J. Donohoe and D. W. Wertz
      J. Phys. Chem. 89 (1985) 282-285

121   R. Rossetti, S. M. Beck and L. E. Brus
      J. Am. Chem. Soc. 105 (1983) 1106-1111

122   W. H. Woodruff, R.F. Dallinger, M.Z. Hoffman, P. G. Bradley, D. Presser
      et.al.
      In ref. 25, 147-155

123   P. G. Bradley, N. Kress, B. A. Hornberger, R. F. Dallinger and W. H.
      Woodruff
      J. Am. Chem. Soc. 103 (1981) 7441-7446

124   R. F. Dallinger, S. Farquharson, W. H. Woodruff and M. A. J. Rodgers
      J. Am. Chem. Soc. 103 (1981) 7433-7440

125   R. F. Dallinger, J. J. Guanchi, W. H. Woodruff and M. A. J. Rodgers
      J. Am. Chem. Soc. 101 (1979) 1355-1357

126   R. F. Dallinger and W. H. Woodruff
      J. Am. Chem. Soc. 101 (1979) 4391-4393

127   N.-H. Jensen and R. Wilbrandt
      In "Photosynthesis I. Photophysical processes – membrane energization",
      G. Akoyunoglou, Ed., Balabin Int. Science Services, Philadelphia,
      Pa, 1981,   97

128   M. Gordon and W. R. Ware (Eds.):
      "The Exciplex", Academic Press, New York 1974

129   S. L. Mattes and S. Farid in "Organic Photochemistry", A. Padwa, Ed.,
      Dekker, New York 1983, Vol. 6, pp. 233-326

130   F. D. Lewis, T.-I. Ho and J. T. Simpson
      J. Org. Chem. 46 (1981) 1077-1082

131   F. D. Lewis and T.-I. Ho
      J. Am. Chem. Soc. 99 (1977) 7991-7996

132   F. D. Lewis
      Acc. Chem. Res. 12 (1979) 152

133   F. D. Lewis, T.-I. Ho and J. T. Simpson
      J. Am. Chem. Soc. 104 (1982) 1924-1929

134   C. Takahashi and S. Maeda
      Chem. Phys. Letters 28 (1974) 22-26

135   L.R. Dosser, J. B. Pallix, G. H. Atkinson, H.C. Wang, G. Levin and M.
      Szwarc
      Chem. Phys. Letters 62 (1979) 555-561

136   G. Levin, B. E. Holloway, C. R. Mao and M. Szwarc
      J. Am. Chem. Soc. 100 (1978) 5841-5844

137   D. Rehm and A. Weller
      Isr. J. Chem. 8 (1970) 259-271

138   J. Eriksen and C. Foote
      J. Phys. Chem. 82 (1978) 2659-2662

139 L. T. Spada and C. S. Foote
     J. Am. Chem. Soc. 102 (1980) 391-393

140 J. Eriksen and C. S. Foote
     J. Am. Chem. Soc. 102 (1980) 6083-6088

141 W. Hub, S. Schneider, F. Dörr , J. D.Oxman and F. D. Lewis
     J. Am. Chem. Soc. 106 (1984) 701-708

142 see ref. 116 and references therein

143 Z. Meić and H.Güsten
     Spectrochim. Acta 34 A (1978) 101-111

144 H. Hamaguchi, T. Urano and M. Tasumi
     Chem. Phys. Letters 106 (1984) 153-156

145 M. Pfeiffer, A. Lau, W. Werncke and L. Holz
     Optics Commun. 363 (1982) 41

146 T. Kamisuki, I. Moriyama, R. Igarashi, Y. Adachi and S. Maeda
     J. Chem. Phys. 73 (1980) 3500

147 H. Koppel, W. Domcke, L. S. Cederbaum and W. von Niessen
     J. Chem. Phys. 69 (1978) 4252-4263

148 D. J. Belville and N. L. Bauld
     J. Am. Chem. Soc. 104 (1982) 294-295

149 N. C. Handy, R. H. Nobes and H.-J. Werner
     Chem. Phys. Letters 110 (1984) 459-463

150 P. D. Mollere, K. N. Houk, D. S. Bomse and T. H. Morton
     J. Am. Chem. Soc. 98 (1976) 4732-4736

151 M. Forster, R. E. Hester, B. Cartling and R. Wilbrandt
     Biophys. J. 38 (1982) 111-116

152 S. Schneider and W. Hub
     In "Proceedings of the 40th Annual Molecular Spectroscopy Symposium",
     Ohio, U.S.A., June 17-21, to be published 1985

153 J. A. Syage, P. M. Felker and A. H. Zewail
     J. Chem. Phys. 81 (1984) 4685-4705

154 S. P. A. Fodor, R. P. Rava, T. R. Hays and T. G. Spiro
     J. Am. Chem. Soc. 107 (1985) 1520-1529

155 L. D. Ziegler and B. Hudson
     J. Chem. Phys. 79 (1983) 1197

156 A. B. Myers and R. A. Mathies
     J. Chem. Phys. 81 (1984) 1552-1558

157 J. M. Friedman and K. B. Lyons
     Nature 284 (1980) 570-572

158 J. M. Friedman, R. A. Stepnoski, M. Stavola, M. R. Ondrias and R. L.
     Cone
     Biochemistry 21 (1982) 2022-2028

159 J. M. Friedman, T. W. Scott, R. A. Stepnoski, M. Ikeda-Saito and T. Yonetani
     J. Biol. Chem. 258 (1983) 10564-10572

160 T. W. Scott, J. M. Friedman, M. Ikeda-Saito and T. Yonetani
     FEBS Letters 158 (1983) 68-72

161 K. B. Lyons and J. M. Friedman
     In "Hemoglobin and oxygen binding", H. Ho, Ed., Elsevier 1982, 333-338

162 M. R. Ondrias, J. M. Friedman and D. L. Rousseau
     Science 220 (1983) 615-617

163 H. Hamaguchi, T. Urano, M. Tasumi, C. Richter and W. Hub
     In preparation

164 P. Stein, J. Terner and T. G. Spiro
     J. Phys. Chem. 86 (1982) 168-170

165 J. Terner, J. D. Stong, T. G. Spiro. M. Nagumo, M. F. Nicol and M. A. El-Sayed
     In "Hemoglobin and oxygen binding", C. Ho, Ed., Elsevier 1982, 355-361

166 W. H. Woodruff and S. Farquharson
     Science 201 (1978) 831-833

167 J. M. Friedman, D. L. Rousseau, M. R. Ondrias and R. A. Stepnoski
     Science 218 (1982) 1244-1246

168 M. J. Irwin and G. H. Atkinson
     In ref. 25, 287-296

169 J. M. Friedman
     In ref. 55

170 D. L. Rousseau
     In ref. 55

171 L. V. Haley and J. A. Koningstein
     J. Phys. Chem. 87 (1983) 621-625

172 L. V. Haley and J. A. Koningstein
     Chem. Phys. 77 (1983) 1-9

173 N.-H. Jensen, R. Wilbrandt, P. B. Pagsberg, A. H. Sillesen and K. B. Hansen
     J. Am. Chem. Soc. 102 (1980) 7441-7444

174 L. V. Haley, A. C. de Wilton, I. W. Wylie and J. A. Koningstein
     In ref. 56, pp. 128-129

175 N.-H. Jensen, R. Wilbrandt, R. V. Bensasson, H.-R. Sliwka and S. Liaaen-Jensen
     In ref. 55

176 H. Hamaguchi
     J. Molec. Struct. 126 (1985) 125-132

177 R. Wilbrandt, N.-H. Jensen and C. Houee-Levin
     Photochem. Photobiol. 41 (1985) 175

178   P. R. Carey
      In ref. 56, pp.70-73

179   P. R. Carey
      In ref. 55

180   Y. Nishina, K. Shiga, R. Miura, H. Toyo, Y. Miyake, T. Yamano, and H. Watari
      In ref. 56, pp. 152-153

181   Y.Nishina, K.Shiga, R.Miura, H.Tojo, M.Ohta, Y.Miyake,T.Yamano and H. Watari
      J. Biochem. 94 (1983) 1979-1990

182   K. Yokoyama and T. Kobayashi
      Chem. Phys. Letters 85 (1982) 175-178

183   W. Hub, S. Schneider and F.Dörr
      In ref. 25, pp. 183-190

184   H. Hamaguchi
      J. Molec. Struct. 126 (1985) 125-132

185   H. Shindo and G. H. Atkinson
      In ref. 25, pp.191-197

186   T. L. Gustafson, D. A. Chernoff, J. F. Palmer and D. M. Roberts
      In ref. 55

187   S. Hirukawa, Y. Torii, S. Suzuki, S. Matsubara and H. Takahashi
      In ref. 56, pp. 136-137

188   N.-S. Lee and M. D. Morris
      In ref. 56, pp. 316-317

189   C. Kato, H. Hamaguchi and M. Tasumi
      In ref. 56, pp. 138-139

190   C. Kato, H. Hamaguchi and M. Tasumi
      J. Phys. Chem. 89 (1985) 407-410

191   M. Braiman and R. Mathies
      Proc. Natl. Acad. Sci. USA 79 (1982) 403 - 407

192   S. O. Smith, M. Braiman and R. Mathies
      In ref. 25, 219-230

193   S.O.Smith, J.A.Pardoen, P.P.J.Mulder, B.Curry, J.Lugtenburg and R. Mathies
      Biochem. 22 (1983) 6141-6148

194   T. Alshuth, I. Grieger and M. Stockburger
      In ref. 25, 231-237

195   A. B. Myers, R. A. Harris and R. A. Mathies
      J. Chem. Phys. 79 (1983) 603-613

196   T. Alshuth and M. Stockburger
      Ber. Bunsenges. Phys. Chem. 85 (1981) 484-489

197   M. Braiman and R. Mathies
      Biochem. 19 (1980) 5421-5428

198 C.-L. Hsieh, M. A. El-Sayed, M. Nicol, M. Nagumo and J.-H. Lee
     Photochem. Photobiol. 38 (1983) 83-94

199 M. A. El-Sayed
     In: "Trends in Photobiology", C. Helene, M. Charlier, T. Montenay-
     Garestier and G. Laustriat, Eds., Plenum Publ. Corp. 1982, 1-10

200 C.-L. Hsieh, M. Nagumo, M. Nicol and M. A. El-Sayed
     J. Phys. Chem. 85 (1981) 2714-2717

201 M. A. El-Sayed and J. Terner
     Photochem. Photobiol. 30 (1979) 125-132

202 J. Terner, C.-L. Hsieh and M. A. El-Sayed
     Biophys. J. 26 (1979) 527-541

203 J. Terner, C.-L. Hsieh, A. R. Burns and M. A. El-Sayed
     Biochem. 18 (1979) 3629-3634

204 A. Champion, M. A. El-Sayed and J. Terner
     In: "Proc. Soc. Photo-Optical Instr. Eng." (SPIE) ,Vol 113, 1977,
     128-135

205 M. Stockburger, T. Alshuth and P. Hildebrandt
     In ref. 55

206 R. A. Mathies
     In ref. 55

207 P. A. Mabrouk and M. S. Wrighton
     Chem. Phys. Letters 103 (1984) 332-335

208 A. Campion, M. A. El-Sayed and J. Terner
     Biophys. J 20 (1977) 369-375

209 A. Campion, J. Terner and M. A. El-Sayed
     Nature 265 (1977) 659-661

210 J. Terner, A. Campion and M. A. El-Sayed
     Proc. Natl. Acad. Sci. USA 74 (1977) 5212-5216

211 G. N. R. Tripathi and R. H. Schuler
     J. Chem. Phys. 76 (1982) 4289-4290

212 S. M. Beck and L. E. Brus
     J. Am. Chem. Soc. 105 (1983) 1106-1111

213 R. E. Hester and K. P. J. Williams
     J. Chem. Soc. Faraday Trans. 2, 77 (1981) 541-547

214 R. Rossetti, S. M. Beck and L. E. Brus
     J. Phys. Chem. 87 (1983) 3058-3061

215 S. M. Beck and L. E. Brus
     J. Am. Chem. Soc. 104 (1982) 4789-4792

216 H. Takahashi, H. Ohkusa, H. Isaka, S. Suzuki and S. Hirukawa
     In ref. 55

217 Y. C. Chung, N. Leventis, P. J. Wagner and G. E. Leroi
     J. Am. Chem. Soc. 107 (1985) 1415-1416

218 S. M. Beck and L. E. Brus
     J. Am. Chem. Soc. 103 (1981) 2495-2496

219 T. Kobayashi, S. Koshihara and T. Hattori
     In ref. 55

220 J. N. Moore, D. Phillips, P. M. Killough and R. E. Hester
     In ref. 55

221 L. V. Haley and J. A. Koningstein
     Chem. Phys. 73 (1982) 263-270

222 S. Suzuki, S. Hirukawa, S. Matsubara and H. Takahashi
     In ref. 56, pp. 318-319

223 J. W. Schindler and J. I. Zink
     J. Am. Chem. Soc. 103 (1981) 5968-5969

224 E. W. Findsen and M. R. Ondrias
     J. Am. Chem. Soc. 106 (1984) 5736-5738

225 T. Uno, Y. Nishimura, and M. Tsuboi
     In ref. 56, pp. 494-495

226 W. H. Woodruff and S. Farquharson
     Anal. Chem. 50 (1978) 1389-1391

227 T. Kitagawa, T. Ogura, and S. Hashimoto
     In ref. 55

228 S. M. Beck and L. E. Brus
     J. Am. Chem. Soc. 104 (1982) 1805-1808

229 B. Robert and M. Lutz
     In ref. 55

230 J. L. Albert, J. Aubard and J. E. Dubois
     J. Raman Spectrosc. 14 (1983) 83-86

231 A. R. Leheny, R. Rossetti and L. E. Brus
     J.Phys.Chem. 89 (1985), 211-213

232 K. Yokoyama
     Chem. Phys. Letters 92 (1982) 93-96

233 M. Asano and J. A. Koningstein
     Can. J. Chem. 60 (1982) 2207-2215

234 M. Asano and J. A. Koningstein
     In ref. 25, 157-160

235 R. Rossetti, S. M. Beck and L. E. Brus
     J. Am. Chem. Soc. 104 (1982) 7322-24

236 J. Terner, C.-L. Hsieh, A. R. Burns and M. A. El-Sayed
     Proc. Natl. Acad. Sci. USA 76 (1979) 3046-3050

237 M. Stockburger, W. Klusmann, H. Gattermann, G. Massig and R. Peters
     Biochem. 18 (1979) 4886-4900

238 Y. Fujimura, Y. Ohtsuki and T. Nakajima
     Bull. Chem. Soc. Japan 58 (1985) 595-600

239   J. M. Friedman
      In ref. 25, pp. 307-315

240   R. E. Hester
      In ref. 25, 177-181

241   M. Asano, L. V. Haley and J. A. Koningstein
      In ref. 25, 139-145

242   I. Grieger, G. Rumbles, and G. H. Atkinson
      In ref. 55

243   G. N. R. Tripathi and R. H. Schuler
      In ref. 25, 199-206

244   W. Hub, C. Richter, H. Hamaguchi, T. Urano and M. Tasumi
      To be published

245   R. Rossetti and L. E. Brus
      J. Am. Chem. Soc. 106 (1984) 4336-4340

246   M. Delhaye, A. Deffontaine, A. Chapput and M. Bridoux
      J. Raman Spectrosc. 15 (1984) 264-267

247   J. L. Goodman and K. S. Peters
      J. Am. Chem. Soc. 107 (1985) 1441-1442

248   H. Hamaguchi, T. Urano, M. Tasumi, C. Richter and W. Hub
      in preparation

249   T. Shida and W. H. Hamill
      J. Chem. Phys. 44 (1966) 2375-2377

250   H. Levanon and P. Neta
      Chem. Phys. Letters 48 (1977) 345-349

251   S. Takeda, K. Kuronayagi, S. Matsubara and H. Takahashi
      In ref. 56, pp. 322-323

Table 1

List of transients[a] studied by TR^3 spectroscopy[§]

Substrate	Transient	Reference

Biology/Biochemistry

==

Substrate	Transient	Reference
Hemoglobin	E	5,26,27,32-34,81,83, 85,157-162,164-170
Cytochrome c	redox	30,47,112,224-227
Bacteriorhodopsin	I	22-24,28,52,75-78, 191-210,242
β-Carotene	E	124,171,174
	S	172
	T	40,118,125,127, 173,175
	I	67
Retinal	I	39
	T	50,113,114,176,177
Enzyme-substrate complex		3,88,89,178,179
Flavin-nucleotide	A	180,181
Reaction centers		74,179,229

Table 1 continued

Organic compounds

===

Aliphatic compounds

1,4-Diaza[2.2.2]bicyclo-octane (DABCO)	C	70,71
Dimethylsulfide	C	119
N-Methylthioacetamide		190

Olefines

Heptatriene	T	115
Polyenes	T	95,115,114
trans-Stilbene	S	49,54,57,58,144,184,186
	A	103,105,135,183,185
	C	37,100,125
	dianion	135
trans-stilbene derivatives	A,C	102
trans-4-nitrostilbenes	T,A	82
5H-Dibenzo(a,d)cycloheptene	A	185
Diphenylbutadiene	S	43,54,58
	T	44

Table 1 continued

Aromatic compounds

Naphthalene	T	46
Anthracene	T	42,45,98
	C	98
Phenanthrene	T	97
Chrysene	S	219
	T	73,96
	C	98
Polyacenes	T	73,95-98
Biphenyl	S,C	189
p-Terphenyl	T	59
	A	60
Phenol	R	90,91,117
Anilin	C	62
N,N-Dimethylaniline	C	31
Phenylacetate	R,I	228
Aminophenol	R	86,211,243
1,4-Dimethoxybenzene	C	29,71
N,N,N',N'-Tetramethyl-p-phenylenediamine (TMPD)	T	232
Methylviologen	C	116,207,240
Diphenylamine	T	48

Table 1 continued

Benzophenone	R	216
Benzoquinone	T	214
	A	65,84,87,92,93,215,243
Anthraquinone-2,6-di-sulphonate	A	41
2,3,4,4-Tetrachlor--1-oxo-naphthalene (β-TKN)	R	106-108
2-(2',4'-Dinitro-benzyl)pyridine (α-DNBP)	I	182,187
2,2',4,4'-Tetranitro-diphenylmethane	I	222
Spiro compounds	I	230 (CARS:101,251)

Heterocycles, Dyes

Acridine	T	45,99
Benzidine	C	188
Flavon	R	220
Methylorange	S	221
Phenazine	T	99
Hydrophenazine	C	110
Phenothiazine	C	111

Table 1 continued

Phenoxazine	C	111
Psoralen	T	188
Quinoxaline	T	51,218

Inorganic compounds

===

Cl_2, Br_2	A	61,63
$M(bpy)_3^{3+}$, M=Cr, Ru	E	35,36,64,83,122,123, 126,217
Nitroprusside	E	223
Rhodanide	A	109
$(UO_2)_3(OH)_5^+$	E	233,234

Surfaces, micelles, colloids, crystals

===

Substrate	Surface	Transient	Reference
Anthracene	SDS	C,T	98
Chrysene	CTAB	C,T	98
Eosin Y	TiO_2	C	245
Methylviologen	TiO_2,CdS	C	121,235
SCN^-	TiO_2	SCN^-,$(SCN)_2^-$	235

Table 1 continued

Tetramethylbenzidine	CTAB	TMB^+, TMB^{++}	212
2,3,4,4-Tetrachlor-1-oxo- naphthalene (β-TKN)	crystal	R	107
Tetrathiafulvalene	Pt	C	231
β-Carotene	crystal	E	174

a   Only transients in fluid media or crystals; no gases,
     flames, low-temperature matrices or stable radicals
     included

§   Abbreviations: S   excited singlet state

                           T   excited triplet state

                           V   excited vibronic state

                           E   excited electronic state

                           A   anion radical

                           C   cation radical

                           R   radical

                           I   isomer

Table 2

Resonance Raman Frequencies and Assignments of Deuterated <u>trans</u>-Stilbene Ion Radicals

ANION RADICALS

	TS neutr	$TS^{-\cdot}$	$DAA^{-\cdot}$	$D12^{-\cdot}$	$D10^{-\cdot}$	$D33^{-\cdot}$	$D44^{-\cdot}$	$DO1^{-\cdot}$	$DAP^{-\cdot}$	$DAP'^{-\cdot}$	$DO5^{-\cdot}$
C=C stretch (e)	1642	1553	1515	1507	1555	1552	1553	1536	1531	1536	1554
C=C stetch ring	1598	1577	1575	1546		1572	1576	1578	1574	1577	1570
CH def (e)	1325	1251	987	990	1243	1246	1249	1225	1223	1222	1243
							975	971	968		
CH def ring	993	978	987	942	943	981	965				972
ring breath	854	848	795				837	818	817	825	824
ring def	640	624	613	587	599	620	619	623	615	618	613

CATION RADICALS

	TS neutr	$TS^{+\cdot}$	$DAA^{+\cdot}$	$D12^{+\cdot}$	$D10^{+\cdot}$	$D33^{+\cdot}$	$D44^{+\cdot}$	$DO1^{+\cdot}$	$DAP^{+\cdot}$	$DAP'^{+\cdot}$	$DO5^{+\cdot}$
C=C stretch (e)	1642	1562	1531	1521	1569	1572	1568	1577	1552	1552	
C=C stretch ring	1598	1605	1596	1567	1592	1604	1602	1600	1596	1597	1600
CH def (e)	1325	1285	1018	1018	1281	1284	1285	1257	1254	1259	1293
CH def ring	1187										
ring def	993	989	992	953		993	977	989	996	990	
ring breath	854	859					859			845	842
ring def	640	628				629	629	624		629	622

Figure Captions

Figure 1: Schematic diagram of an experimental setup for
picosecond RR studies using a single laser /27/.

Figure 2: (a) RR spectra of COHb with 30 ps pulses (576 nm)
in perpendicular and parallel polarizations. Spectra
(a,d) of the partially photolyzed sample are measured
with the tightly focussed laser beam, the spectra
(b,e) of unphotolyzed COHb are obtained with diffuse
focus. RR spectra of phototransients (e,f) are achieved
by proper subtraction. The subtraction c-f yields
spectrum g which shows the position of band IV at
1552 cm^{-1} being characteristic of high-spin Fe II
in a planar heme /27/.

(b) RR spectra of COHb with 20-ns pulses (576 nm)
in perpendicular and parallel polarization: Subtractions
as in the left parts /27/.

Figure 3: Schematic diagram of a nanosecond pump-probe TR[3]
apparatus using a nitrogen laser for photolysis
and a excimer laser pumped dye laser for Raman
probing /37/. Multichannel detection is achieved with
a gated 4-stage image intensifier and a vidicon
tube. 180° backscattering geometry is used at the
sample.

Figure 4: Schematic diagram of a picosecond pump-probe TR^3
apparatus with two tunable dye lasers and multichannel detection /58/. ML mode locker; SPDL synchronously pumped dye laser; CD cavity dumper; PR
polarization rotator; SHG second harmonic generator;
BS dichroic beam splitter; RP retroprism; S sample;
F filter; SPECT spectrograph; OMA optical multichannel analyzer. The scattered light is collected
at 90° from the sample.

Figure 5: Schematic diagram of a TR^3 apparatus for radiation-
chemical studies /7/. The electron accelerator deli-
vers 30 ns to 5 µs electron pulses of 2-2,5 MeV
energy and 1,2 A current at a repetition rate of
up to 60 Hz.

Figure 6: Bacteriorhodopsin photocycle showing the main features of the C=N configurations of the N-protonated
intermediates /52/.

Figure 7: (a) RR spectrum of BR_{548} in H_2O(A) and D_2O(B);
$[14,15-^{13}C_2]BR_{548}$ in H_2O(C) and D_2O (D);$[12,14-D_2]BR_{548}$
in H_2O(E) and D_2O(F).
(b) Similar conditions for RR spectra of L_{550}/52/.
From the deuteration-induced shifts in the fingerprint
region the authors conclude on the C=N configuration
to be cis in BR_{548} and trans in L_{550}.

Figure 8: (a) Time-evolution of the low-frequency RR spectrum of the deoxy photoproduct from COHb. After 9 ns a band at 345 cm^{-1} band is almost fully developed. This band is characteristic for a R-state configuration as it is also observed in a mutant human hemoglobin (Kempsey; lower spectrum) which is known to be stabilized in the R structure at high pH even when ligand free /33/.

(b) Time-evolution of v_{Fe-His} of human deoxy Hb after photolysis. The corresponding Raman band from the equilibrium deoxy HbA is shown in the bottom. Whereas the R-state structure is fully achieved within 9 ns after photolysis (see Figure 8a) the Fe-His frequency still has to relax to its equilibrium value at 215 cm^{-1} /33/.

Figure 9: (a) Raman Spectrum of trans-stilbene (0.03 M) and ethyldiisopropylamine (0.75 M) in acetonitril (λ=480 nm); (b) at 60 ns after photolysis with a nitrogen laser pulse; (c) difference spectrum containing the RR spectrum of TS$^{\bar{}}$ /103/.

Figure 10: Second-order decay of TS$^{\bar{}}$ monitored by the intensity of the 1577/1553 cm^{-1} RR band in the presence of various amines /103/.

Figure 11:RR spectrum of the photointermediate in an acetonitrile solution of <u>trans</u>-stilbene (0.05 M) in the
presence of (a) fumaronitrile (0.2 M) or (b) 9,10-
dicyanoanthracene (4×10^{-4}M) at 60 resp. 80 ns after
the photolysis pulse /37/.

Figure 12:Deuterated <u>trans</u>-stilbenes whose anion and cation
radicals have been measured by TR^3 spectroscopy
/100/.

Figure 13:RR spectra of normal and deuterated <u>trans</u>-stilbene
anion radicals.

Figure 14:RR spectra of normal and deuterated <u>trans</u>-stilbene
cation radicals.

Scheme 2: Mechanism of reaction of singlet excited <u>trans</u>-
stilbene with tertiary amines in acetonitrile solution /103/.

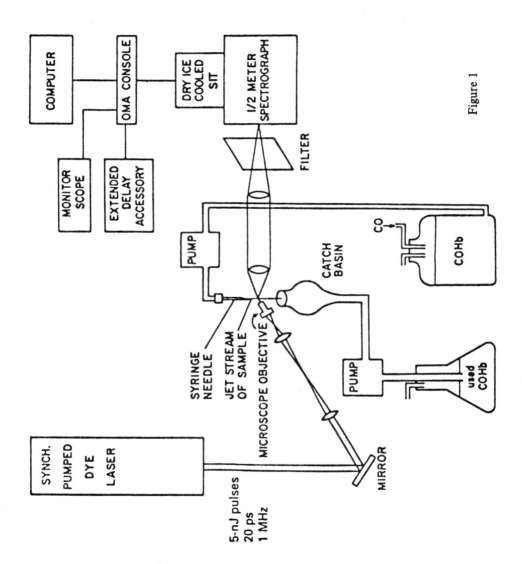

Figure 1

320

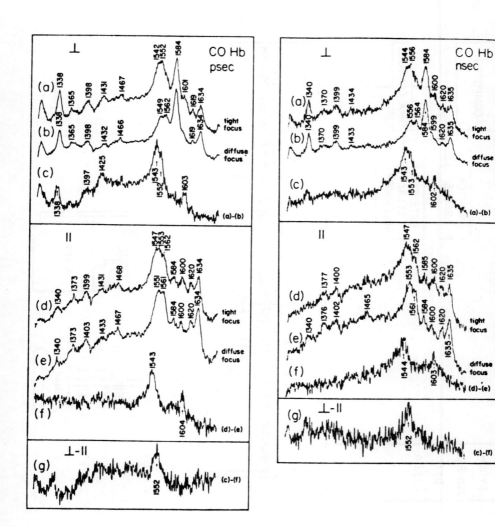

Figure 2

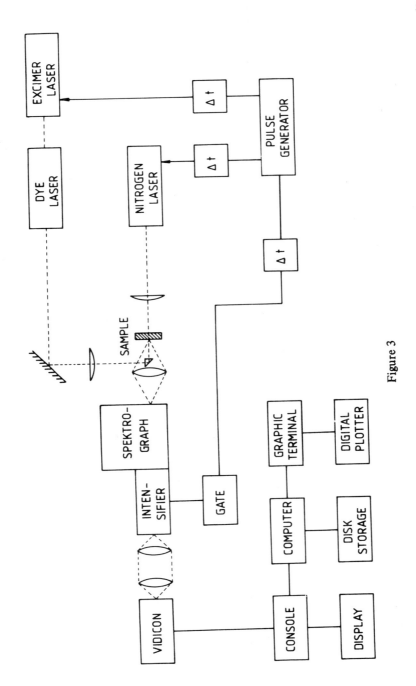

Figure 3

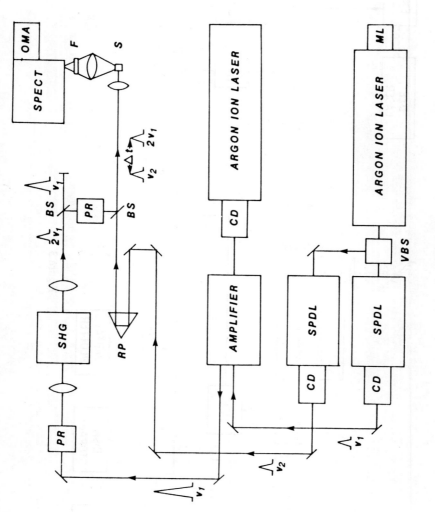

Figure 4

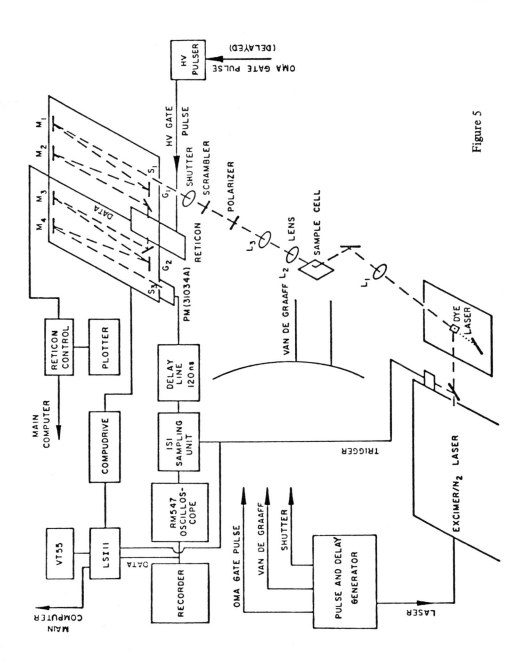

Figure 5

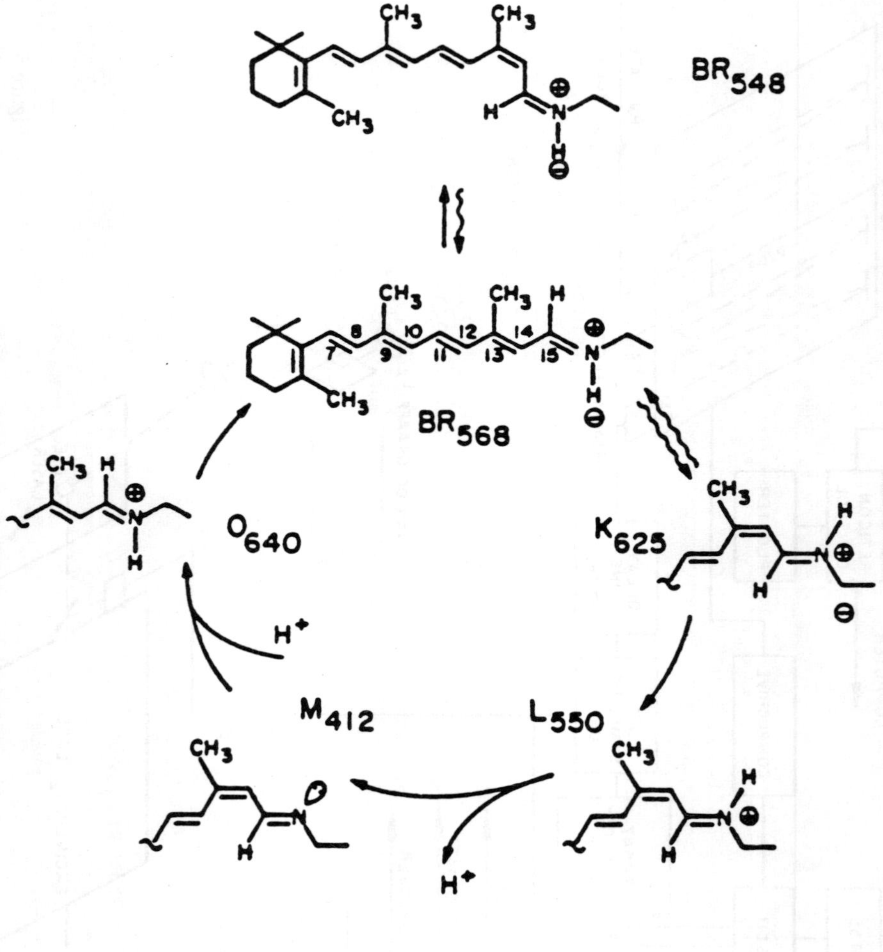

Figure 6

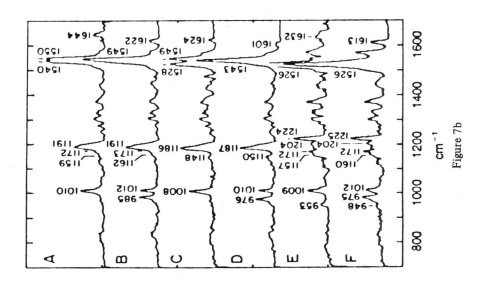

Figure 7b

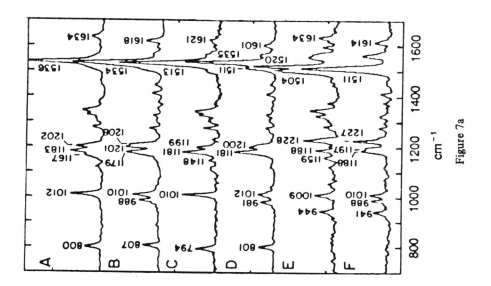

Figure 7a

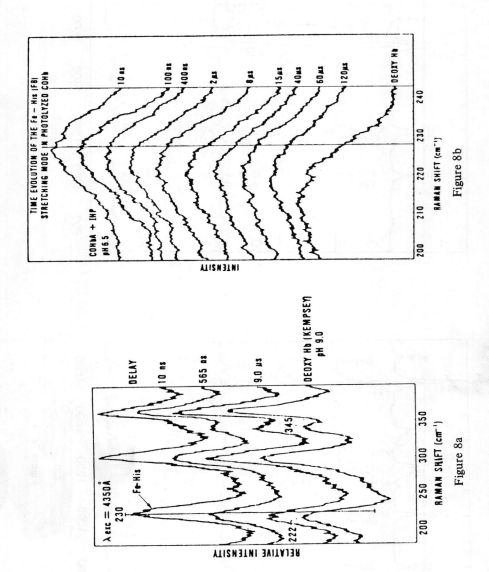

Figure 8b

TIME EVOLUTION OF THE Fe — His (F8)
STRETCHING MODE IN PHOTOLYZED COHb

COHbA + IHP
pH 6.5

10 ns
100 ns
400 ns
2 µs
8 µs
15 µs
40 µs
60 µs
120 µs
DEOXY Hb

INTENSITY

RAMAN SHIFT (cm⁻¹)

Figure 8a

DELAY
10 ns
565 ns
9.0 µs
DEOXY Hb (KEMPSEY)
pH 9.0

λ exc = 4350 Å
230
Fe-His
222
345

RELATIVE INTENSITY

RAMAN SHIFT (cm⁻¹)

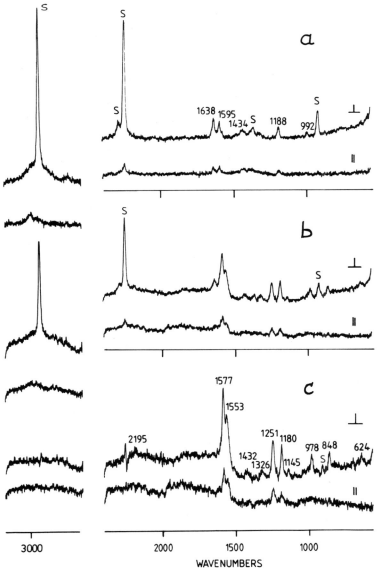

Figure 9

328

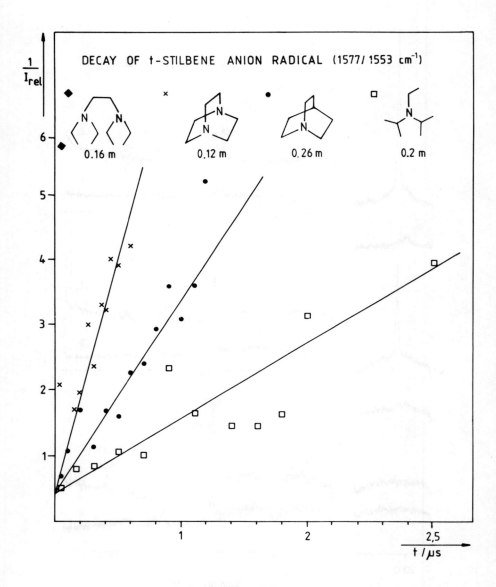

Figure 10

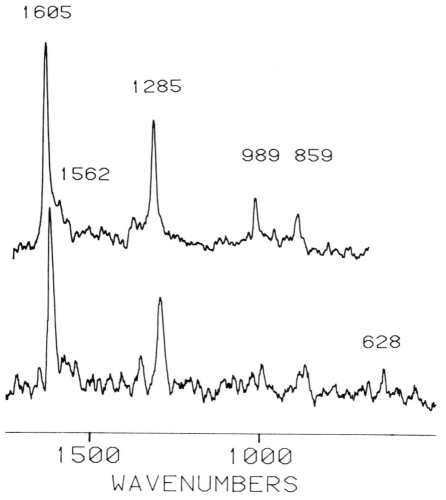

Figure 11

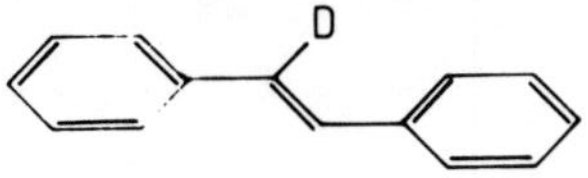

Figure 12

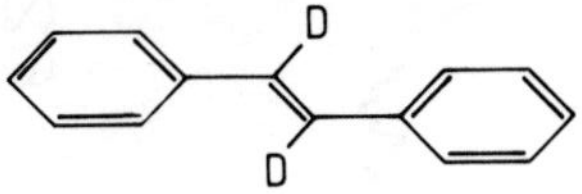

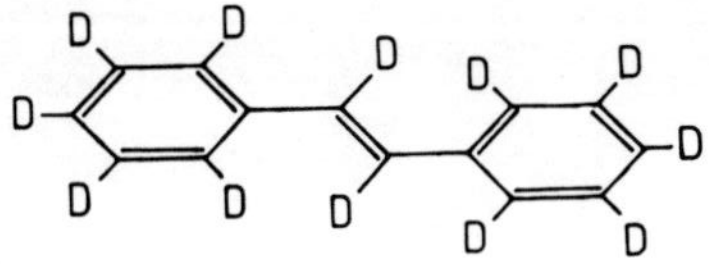

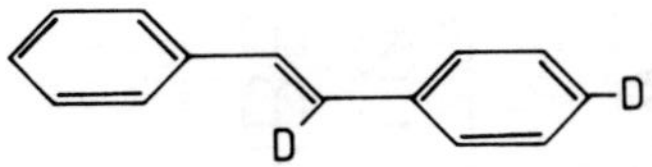

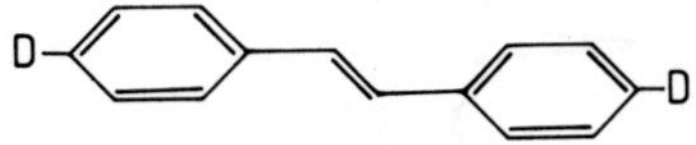

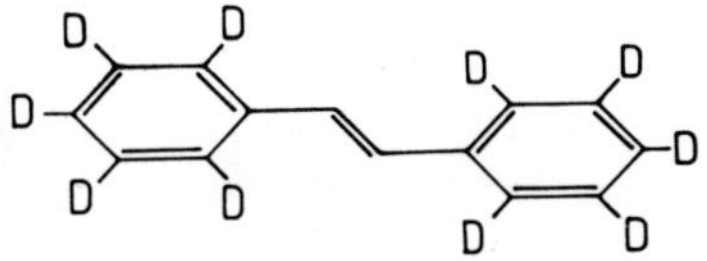

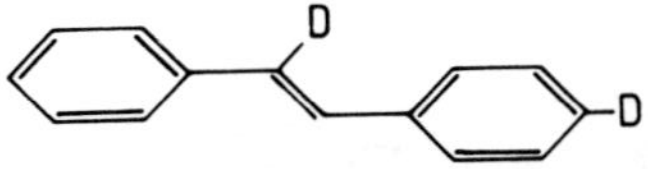

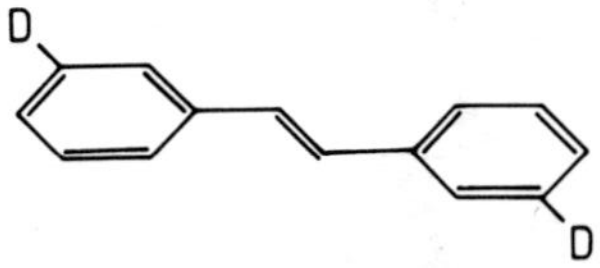

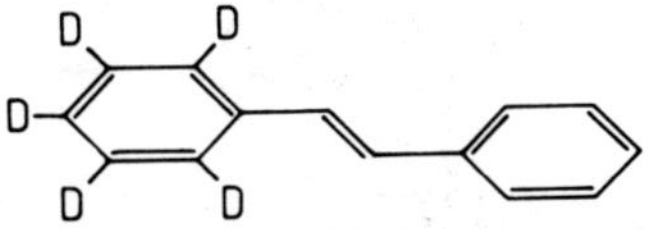

331

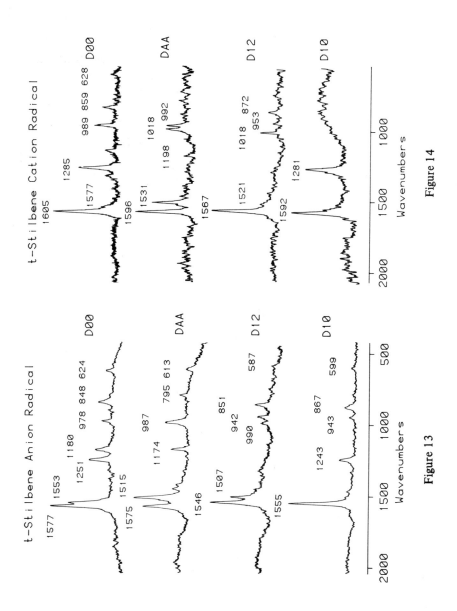

Figure 14

Figure 13

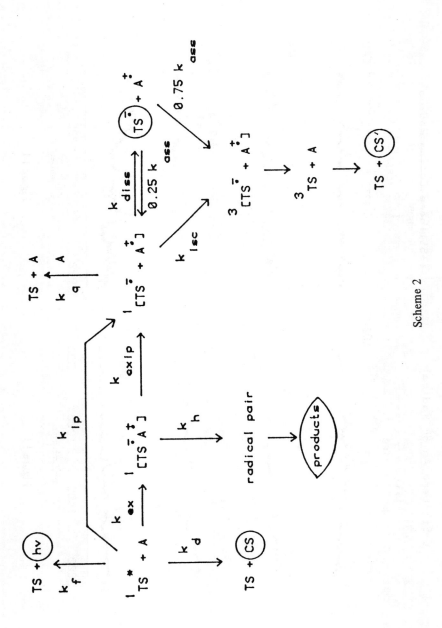

Scheme 2

5. DETECTION AND SPECTROSCOPY OF METHYL AND SUBSTITUTED METHYL RADICALS BY RESONANCE ENHANCED MULTIPHOTON IONIZATION

M. C. Lin

Chemistry Division
Code 6105
Naval Research Laboratory
Washington, D. C. 20375-5000, USA

and

W. A. Sanders

Department of Chemistry
The Catholic University of America
Washington, D. C. 20064-0001, USA

334

Contents

I. Introduction

In recent years intense, narrow-linewidth tunable dye lasers have become available over a steadily expanding range of the spectrum. This has greatly improved our ability to detect free radicals at very low concentrations (typically $\lesssim 10^9$ particles/cc) and to study their spectroscopy and reaction kinetics. One of the most convenient and popular detection methods employed to date is laser-induced fluorescence (LIF) [1-3]. The high sensitivity and spatial resolution of LIF make it possible to probe species of low concentration in very small volumes. By contrast, resonance absorption (RA) typically requires a long absorption path to obtain measurable attenuation. This disadvantage can be partially overcome by using multiple reflections, but RA still requires a much larger active volume than LIF. On the other hand, RA can be employed to detect "dark" species which do not fluoresce from excited states due to rapid predissociation or internal conversion. The species which do not fluoresce are at least as abundant as those which do.

More recently, the new technique of multiphoton ionization (MPI) and its more sensitive variation utilizing the effect of resonance enhancement (REMPI) have been developed. REMPI combines the advantages of LIF and RA, being capable of detecting both "dark" and fluorescing species in a tightly focused volume with high sensitivity and selectivity. It is a particularly versatile technique if employed in conjunction with a mass spectrometer to resolve the masses of the ionized species.

The REMPI method and its application to studies of molecular spectroscopy have been subjects of much current interest and discussion [4,5]. In this brief account, we will describe the application of REMPI/MS to the detection and spectroscopy of nonfluorescing free radicals. The discussion will focus on the results for CH_3 and substituted methyl radicals obtained recently at NRL, where most of the published work on radicals has been done.

Our first attempt to employ MPI for the detection of nonfluorescing radicals was made in 1979, using a discharge flow system [6]. At least two reactions,

$$O(^3P) + C_2H_2 \longrightarrow CH_2(^3B_2) + CO$$

$$O(^3P) + C_2H_4 \longrightarrow CH_3 + CHO \; ,$$

were utilized as sources of CH_2 and CH_3 radicals [7]. This experiment was unsuccessful, due in part to the low power of the tunable dye laser employed and possibly to the low concentration of radicals produced by the reactions noted above. These reactions have now been shown to generate other radical species (such as C_2H_3O from $O + C_2H_4$) with relatively large yields [8,9]. Subsequent experiments employing a more powerful pulsed dye laser in conjunction with a quadrupole mass spectrometer proved to be successful in detecting CH_3 [10,11] and several other nonfluorescing radicals. Only CH_3 and its substituted derivatives will be reviewed in detail below.

Much of this review is devoted to the methyl radical because of its great experimental and theoretical importance. It figures prominently in the early stages of hydrocarbon combustion and in atmospheric chemistry. Methyl radicals have been generated for many years by the pyrolysis or photolysis of a variety of precursors, and there is a vast literature on the kinetics of their reactions with many different substrates. Furthermore, methyl radical serves as a benchmark for theoretical calculations because it is one of the simplest polyatomic species of broad experimental interest.

Although methyl radicals have been studied intensively since they were first discovered by Paneth and Hofeditz in 1929 [12], their spectra were unknown until pioneering observations were made by Herzberg nearly thirty years later [13]. He identified three Rydberg series in the vacuum ultraviolet and was able to resolve some vibronic structure in both CH_3 and CD_3. Analysis of these spectra indicated that the methyl radical must be planar with an uncertainty of less than 10°.

More recently, additional evidence has come from photoelectron spectra [14,15], photoelectron detachment spectra [16], electron spin resonance [17,18], and both gas-phase [19-21] and matrix-isolation [22-25] infrared spectra. In all cases, interpretation of the observed spectra required the assumption of planarity. Very convincing evidence was provided by the work of Yamada, et al. [20], who performed a high-resolution absorption study using a laser diode spectrometer. By carefully measuring the spacings and relative intensities of absorption bands due to out-of-plane bending, they were able to show that the observed spectra were only consistent with the assumption of D_{3h} symmetry.

The conclusions based on interpretation of spectra are supported by the results of ab initio molecular orbital calculations [26-31], which indicate that the most stable nuclear configuration corresponds to D_{3h} symmetry. Further, there appears to be no inversion barrier. The ground state electron configuration is now generally accepted to be $(1a_1')^2(2a_1')^2(2e')^4(2a_2'')^1$, with the state designation $\tilde{X}\,^2A_2''$.

Figure 1 is a schematic molecular orbital diagram for CH_3, showing the correlation with the corresponding united atom (F). The excited orbitals that can be populated by single-photon transitions from the singly-occupied

a_2'' orbital of the ground state electron configuration are indicated by solid lines. The dashed lines correspond to orbitals that are accessible only through multiphoton transitions. The derivation of single- and multiphoton selection rules for CH_3 is discussed in the following section.

II. Selection Rules for 1-3 Photon Excitations of D_{3h} Species

This section presents a simplified discussion of the multiphoton selection rules involved in the interpretation of the CH_3/CD_3 REMPI spectra. No attempt will be made to derive selection rules for the general case. Instead, the procedure will be outlined for the specific case of interest--namely transitions from an A_2'' electronic ground state of a radical belonging to D_{3h} symmetry. The intent is to provide sufficient detail so that the approach can be extended to other cases by analogy when the need arises. Since most radicals of interest are of lower symmetry than CH_3, their selection rules will generally be less complex than those considered here.

A rigorous derivation of selection rules for simultaneous multiphoton transitions involves defining the conditions under which the transition matrix elements coupling the initial and final states are nonzero. The wavefunctions depend upon both the electronic and nuclear coordinates, and the selection rules are determined by their transformation properties under the symmetry operations of the point group to which the molecule (radical) belongs. In the general case, the interaction between the radiation field and the material system is described by a tensor operator. More specifically, in the dipole approximation the interaction operator for an n-photon process is an nth-order reducible tensor with 3^n components. In order to derive the selection rules in the most general manner, it is necessary to resolve this operator into its irreducible tensor components and to classify them according to the irreducible representations of the point group. If this rather formidable task can be accomplished, then the allowed transitions are simply those for which the direct product of the irreducible representations of the initial and final state wavefunctions contains the representation of one of the components of the interaction operator.

Since most readers are probably not familiar with the manipulation of higher-order tensors, we will not attempt a rigorous derivation of the selection rules in this review. Instead, we will present a simpler heuristic procedure which is adequate for the purpose and hopefully makes up in transparency what it may lack in rigor. We will also make a number of simplifying assumptions that are justified by the nature of the processes of interest. Perhaps most important of these is the assumption that the radicals are vibrationally cold, so that the vibrational contributions to the initial and final state wavefunctions can be ignored. We will further assume that the Born-Oppenheimer approximation is valid; that is, that the state functions can be written as products of an electronic function (electronic coordinates) and a rotational function (nuclear coordinates). We will then proceed to deduce the selection rules for the electronic and rotational quantum numbers separately, taking the coupling between them into account only when it affects the selection rules directly.

As noted above, both spectroscopic evidence and _ab initio_ calculations indicate that CH_3 is a planar symmetric top, belonging to the D_{3h} point group. This means that the electronic states must belong to the irreducible representations of this point group, four of which are nondegenerate (A_1', A_1'', A_2', A_2'') while the remaining two are doubly degenerate (E' and E"). For the present application it is assumed that the initial state is the electronic ground state, which belongs to A_2''. Although it may be convenient to visualize the electronic wavefunction in a specific form, such as an antisymmetrized product (Slater determinant) of suitably optimized molecular symmetry orbitals, the present discussion assumes only that the symmetry properties of the initial and final states are known. Furthermore, all possible excited state symmetries will be considered, regardless of whether they can be constructed from a given finite basis set. In other words, the arguments used are independent of the model used to describe the initial and final state electronic wavefunctions.

In the present case, the correlation problem is considerably simplified by the assumption that the transitions of interest involve Rydberg states resulting from the excitation of the single unpaired electron in the highest occupied orbital in the electronic ground state (a_2''). Thus the symmetry of the overall electronic state is that of the highest occupied molecular orbital. For example, if this orbital is of a_1' symmetry, the state is labeled A_1'.

In our discussion we will assume that single-photon processes can be treated in the dipole approximation. That is, the higher-order terms in the operator describing the interaction between the radical and the electromagnetic field are neglected. The electronic selection rule is then reduced to a rather simple statement: a given transition is allowed only if the direct product of the irreducible representations of the initial and final states contains the representation of one of the components of the dipole operator. If the transition is allowed for the z-component of the dipole operator, it is called a "parallel" transition. On the other hand, if it is allowed for the x- and y-components it is "perpendicular." Since z belongs to A_2'' and (x,y) to E', this means that the direct product of the initial and final state symmetries must contain one of these representations. The irreducible components of the direct products can be obtained directly from tables in Herzberg's Volume III [32].

For single-photon excitation of the $^2A_2''$ ground state of CH_3, only two possible excited state symmetries yield nonzero transition moments according to the criteria discussed above. As indicated in Table I, these are $^2A_1'$ ("parallel") and $^2E''$ ("perpendicular"), corresponding to the direct products $A_2'' \otimes A_1' = A_2''$ and $A_2'' \otimes E'' = E'$.

Rather than attempt to deal specifically with multiphoton operators, we will adopt the expedient of considering multiphoton transitions as sequences of single-photon processes, starting from the initial state and proceeding through one or more intermediate states to the final state. By virtue of the closure property of quantum state vectors, this is a mathematically legitimate procedure for calculating transition moments provided the contributions from all possible intermediate states are included. For the purpose of deducing selection rules, however, it is only necessary to demonstrate that there is at least one accessible intermediate that is capable of providing an allowed path at each step up the ladder.

In order to determine the allowed two-photon transitions from the ground state of methyl radical, therefore, it is only necessary to find those states which are allowed by single-photon transitions from the $^2A_1'$ and $^2E''$ states. The results are shown in Table I. In principle, this inelegant but conceptually simple process can be continued indefinitely to arrive at the electronic selection rules for any number of photons. The number of accessible states increases at each step as new combinations become available. For CH_3, all possible symmetry species of D_{3h} become allowed by the time three-photon processes are considered, and most of the transitions are possible by two or more specific paths.

In the most general case, it would be necessary to consider the transformation properties of the vibrational portions of the wavefunctions of the absorbing species under the symmetry operations of the overall point group. Since vibrations are assumed to be unexcited in the work discussed here, however, the vibrational selection rules will not be considered in this discussion.

The equilibrium nuclear geometry of CH_3 is an oblate symmetric top in the rigid rotator approximation. The two moment of inertia components perpendicular to the principal symmetry axis (which in turn is perpendicular to the plane of the radical) are equal. The rotational states are characterized by the total angular momentum quantum number N and its projection K along the (molecule-fixed) principal symmetry axis. Since the rotational energy depends on K^2, there is a two-fold degeneracy corresponding to the two values $\pm|K|$. In addition, there is a (2N+1)-fold degeneracy which reflects the number of possible projections of the total angular momentum vector on the space-fixed z-axis.

The well-known selection rules for rotational transitions of a symmetric top accompanying single-photon electronic transitions are [32]

For parallel bands,

$$\Delta K = 0, \quad \begin{cases} \Delta N = 0, \pm 1 \text{ if } K \neq 0 \\ \Delta N = \pm 1 \text{ if } K = 0 \end{cases}$$

and for perpendicular bands,

$$\Delta K = \pm 1, \quad \Delta N = 0, \pm 1$$

For the case K=0, the rotational wavefunctions reduce to the ordinary spherical harmonics, and the forbidden nature of $\Delta N=0$ is a consequence of the odd parity of the one-photon interaction operator. This result follows mathematically from the symmetry properties of the angular momentum coupling (Clebsch-Gordan) coefficients.

When a multiphoton transition in CH_3 involves one or more of the degenerate (E' or E") electronic states, it is necessary to consider an additional restriction due to the interaction of rotational and electronic motions (coriolis coupling). Each degenerate state is split into two levels (designated +l and -l), and the selection rules must take into account the possible changes of l. Mills [33] has developed a simple recipe for predicting these rules, which for a D_{3h} species are

$$\Delta K = 0: \quad A \longleftrightarrow A, \quad E(+l) \longleftrightarrow E(+l), \quad E(-l) \longleftrightarrow E(-l)$$
$$\Delta K = +1: \quad E(+l) \longleftarrow A, \quad E(-l) \longleftarrow E(+l), \quad A \longleftarrow E(-l)$$
$$\Delta K = -1: \quad E(-l) \longleftarrow A, \quad E(+l) \longleftarrow E(-l), \quad A \longleftarrow E(+l)$$

If there are no degenerate states involved in a given multiphoton transition, these rules introduce no new restrictions. Otherwise, they impose conditions on the allowable changes of the quantum number K. To illustrate the procedure, we consider a specific example of a three-photon transition. The overall transition $A_1'' \longleftarrow A_2''$ is allowed by the path $A_2'' \longrightarrow E'' \longrightarrow E'' \longrightarrow$

A_1''. When the coriolis selection rules are taken into account, the possible specific paths are

$$A_2'' \begin{array}{c} \xrightarrow{+1} E''(+1) \xrightarrow{\;+1\;} E''(-1) \xrightarrow{+1} \\ \xrightarrow{-1} E''(-1) \xrightarrow[-1]{} E''(-1) \xrightarrow{-1} \end{array} A_1'' \qquad \Delta K = \pm 3 \;,$$

where the allowed change of K is indicated above or below the arrow for each step.

For multiphoton transitions between nondegenerate states, there are restrictions on the allowed values of ΔN when a state with K=0 is involved. As mentioned above, for K=0 there is no component of angular momentum along the molecular axis and the symmetric top rotational wavefunction reduces to the ordinary spherical harmonic Y_{NM}. The selection rules can be worked out by considering the transformation properties of the rotational wavefunctions and interaction operators under the group symmetry operations. In the general case, however, this can be a rather complex procedure requiring facility in manipulating the rotation matrices and Clebsch-Gordan coefficients. As an alternative, we will present the following simplified argument.

The direct products of the A_2'' ground state representation with the nondegenerate representations are [32]

$$A_2'' \otimes A_1' = A_2''$$
$$A_2'' \otimes A_1'' = A_2'$$
$$A_2'' \otimes A_2' = A_1''$$
$$A_2'' \otimes A_2'' = A_1'$$

The two A_2 direct-product representations correspond to interaction operators of odd parity. For example, for a single-photon transition R_z belongs to A_2' and z belongs to A_2''. This implies that the product of the rotational wavefunctions must be of odd parity to yield overall even parity, which in turn means that the initial and final values of N must differ by an odd integer. For the two A_2 final states, on the other hand, the direct product is totally symmetric. From this we deduce that the interaction operator is of even parity and hence that $\triangle N$ must be even. The consequences of this final restriction can be seen in Table I.

III. Experimental Methods

The REMPI/MS setup employed by DiGiuseppe et al. [10,11] consisted of three major sections--the radical source, the laser ionization region, and the mass spectrometer ion detector. A schematic diagram of the apparatus is shown in Figure 2.

Various methods have been employed to generate radicals in this laboratory. These include photolysis with pulsed UV (266 nm) [34] and IR (10 μm) [35] lasers, pyrolysis [10,11], and a discharge-flow method utilizing fast atom-molecule reactions [36]. The last two methods are the most convenient and reliable, generating vibrationally colder and cleaner radicals. They will be discussed here.

In the pyrolytic method, methyl radicals were generated by the thermal decomposition of appropriate sources listed in Table II. The deuterated methyl radical, CD_3, was produced mainly from CD_3SOCD_3 (DMSO-d_6). These compounds are known to produce CH_3 radicals cleanly at relatively low temperatures under low pressure conditions, with no spectroscopically complicating fragments being generated within the spectral region of interest. The oven consisted of a resistively-heated crinkly tantalum foil attached to the inner end of a 6-mm stainless steel tube (see Figure 2). The details of construction of this tantalum oven reactor have been described in reference 11. The temperature of the oven was typically maintained at about 1100 K. As indicated by the rate constants given in Table II, the half-lives of the radical source molecules mentioned above are much shorter than 1 msec at this temperature.

Using the discharge-flow method, the methyl and substituted methyl radicals were produced by fast halogen-atom abstraction reactions at room temperature [36]:

$$X + CH_3R \longrightarrow HX + \,^{\bullet}CH_2R \, ,$$

where X = Cl, F and R = H, F, OH, C_2H_3, C_3H_5. For CH_2F, the reaction F + $CH_2CO \longrightarrow CH_2F$ + CO was also employed as the radical source. The halogen atoms were generated by microwave discharge of the halogen molecules upstream of the flow reactor. A schematic diagram of the discharge-flow reactor and the mass spectrometric detection system is shown in Figure 3. The discharge-flow tube was constructed according to the design described by Anderson and Bauer [37], using a 1" O.D. Pyrex tube with a skimmer attached to its end. The skimmer had an orifice of about 700 μm which allowed the radical products of interest to expand into the ionization region of the mass spectrometer with the aid of He carrier gas. The flow velocity was maintained at approximately 1200 cm/sec at 2 torr He pressure. Typically, several mtorr of the radical source molecules (i.e. X_2 and RCH_3) were used to generate the radicals.

As shown in Figures 2 and 3, the ionization of the radicals was achieved by focusing the output of a tunable dye laser into the ionization region of a quadrupole mass spectrometer (Extranuclear Laboratories) by means of an f/2 quartz lens (focal length 50 mm) placed inside the mass spectrometer. The total pressure in the mass spectrometer was typically about 3×10^{-7} torr. The number density of radicals in the ionization region was estimated to be about 10^{10} particles/cc.

The ionized radicals were extracted and focused into the quadrupole mass spectrometer. Either total ions or mass-selected specific ions could be measured. Signals from the ion multiplier (Channeltron) were amplified and pulse-stretched by means of a current multiplier (Keithly 427) for further integration and baseline subtraction. Signal averaging and data processing were aided by a PDP 8E computer, which scanned the dye laser stepwise and collected the ion signal at each step. Typically the signals generated by 10 laser pulses were accumulated and averaged at each wavelength step.

The dye laser, pumped by the second (532 nm) or third (355 nm) harmonic of a Nd:YAG laser (Quantel), was operated optimally at 10 Hz over a broad range of wavelengths of interest for the present application (250-500 nm). A variety of dyes, pumping wavelengths, and frequency-doubling crystals were employed in order to obtain coverage of the entire range (see Table III). The laser output was typically a few mJ in the UV region (250-350 nm) and 10-15 mJ in the visible (350-500 nm).

IV. REMPI Detection and Spectroscopy of Methyl and Substituted Methyl
Radicals

The REMPI spectral data for the methyl and substituted methyl radicals
obtained in this laboratory by Hudgens, DiGiuseppe, Dulcey, and others will
be summarized more or less chronologically in this section. Data on CH_3
reported by researchers from other laboratories will also be included in the
discussion.

A. Methyl (CH_3 and CD_3) Radicals

1. Three-photon resonance enhanced ionization

The first reported detection of CH_3 radicals by MPI was achieved on the
basis of 3-photon resonance with the $\tilde{C}$ $^2E''$ <--- $\tilde{X}$ $^2A_2''$ ($\tilde{\gamma}_I$ Rydberg state)
transition [10] using azomethane (AZM) as the radical source. Figure 4a
shows the effect of oven temperature on the appearance of 450-nm laser
ionization using AZM. The similar but slightly cleaner spectrum shown in
Figure 4b was obtained when DMSO was employed as the CH_3 source. The
continuous DMSO spectrum observed at $25°C$ between 444 and 454 nm is
consistent with the fact that the single-photon spectrum of DMSO in the 150
nm region is continuous [38]. Both spectra shown in Figure 4 clearly
indicate the presence of a strong resonance at 450.8 nm, which was assigned
to $\tilde{\gamma}_I(0,0)$. These spectra were obtained when the oven was heated to $800°C$,
at which temperature CH_3 radicals were produced efficiently from the
unimolecular decomposition of the source compounds. In Figure 4b a small
peak is clearly discernable at 445.8 nm, shifted by 746 cm^{-1} from $\tilde{\gamma}_I(0,0)$.
It was attributed to the $\tilde{\gamma}_I(1,1)$ band of the $\tilde{C}$ <--- $\tilde{X}$ Rydberg transition.

Further scanning toward shorter wavelengths confirmed the 445.8 nm peak and also revealed another at 432.9 nm, which was attributed to $\gamma_1(2,0)$ in preference to the $\beta_1(0,0)$ band of the $\tilde{B}\ ^2A_1'$ <--- $\tilde{X}\ ^2A_2''$ Rydberg transition. The latter requires a two-photon resonance plus an additional two photons for excitation to the continuum. This assignment is partly supported by the wavelength shift of 2750 cm^{-1} from $\gamma_1(0,0)$ to $\gamma_1(2,0)$ which was reported previously by Herzberg [13]. Further corroboration was obtained from the spectral data of CD_3 REMPI using DMSO-d_6 as the radical source. A set of representative spectra for the 430-454 nm region for both CH_3 and CD_3 is reproduced in Figure 5. A similar spectrum for CH_3 was also obtained using di-t-butyl peroxide (DTBP) as the CH_3 source [10].

It should be mentioned that wavelength-resolved spectra for other masses, including m/z = 45, 47, and 63 for DMSO-d_6 and m/z = 43 for AMZ, were also obtained and examined. However, these spectral data did not reveal any distinct features similar to those of the m/z = 15 CH_3 REMPI spectra shown above.

Additional three-photon REMPI transitions of CH_3 and CD_3 were detected in the spectral region between 415 and 435 nm as indicated in Figure 6. The intensities of these higher-energy Rydberg states are obviously lower than those of the γ_1 series. On the basis of the intensity of the $\gamma_1(2,0)$ band, which appeared in both spectral regions represented in Figures 5 and 6, the intensities of these higher transitions were found to be lower by as much as a factor of 75. The small peak at 428.4 nm in the CH_3 spectrum of Figure 6a has been attributed to $\gamma_1(3,1)$, which unfortunately could not be resolved in the analogous CD_3 spectrum [11]. Two other peaks appearing at 421.9 and 415.9 nm in Figure 6a were ascribed to $\beta_2(0,0)$ and $\gamma_2(0,0)$, corresponding to the single-photon absorption wavelengths of 140.7 and 138.5 nm, respectively, reported by Herzberg [13]. The corresponding transitions for CD_3 were also found to be consistent with those given by Herzberg.

All of the assignments for the spectral data obtained by the three-photon REMPI of CH_3 and CD_3 are collected in Table IV for comparison with Herzberg's single-photon VUV absorption data. It is seen that the agreement between the 3-photon MPI data in the UV-visible region and Herzberg's 1-photon absrption data in the VUV region is very good within the uncertainty of the present measurements (± 0.1 nm) due to the diffuseness of the REMPI peaks.

It is evident from the spectra shown in Figures 5 and 6 that the REMPI peaks are derived from both the principal Rydberg state and vibronic state transitions. From the shifted vibronic peaks, the spacings between various vibronic levels can be calculated. For example, the spacing between $\gamma_i(0,0)$ and $\gamma_i(1,1)$ transitions provides the measurement for $\nu_2' - \nu_2'' = 746$ cm^{-1} (in vacuum), which is the difference between the out-of-plane bending frequencies in the upper Rydberg and ground states. Similarly, other frequency differences can be readily computed from the various peaks assigned previously. These data are summarized in Table V. The calculated vibronic spacings are estimated to be accurate to ± 40 cm^{-1} because of the diffuseness of the REMPI transitions.

Using the published value of the ground state vibrational frequency for the out-of-plane bending, $\nu_2'' = 603$ cm^{-1}, and the calculated frequency difference, $\nu_2' - \nu_2'' = 746$ cm^{-1}, we obtained $\nu_2' = 1349$ cm^{-1} [11]. This result and others calculated for high vibronic transitions of both CH_3 and CD_3 are given in Table VI. The analysis of these vibronic transitions was carried out by assuming Herzberg's case b [32]. The results presented in this table for $\Delta v_{2}' = 0\text{-->}1$ and $1\text{-->}2$ agree closely with those obtained by Herzberg [13]. The large increase in the observed values of ν_{2}' of the $\tilde{C}\ ^{2}E''$ Rydberg state (from 603 to 1349 cm^{-1} for CH_{3} and 453 to 1062 cm^{-1} for CD_{3}) seems to be consistent with the results of Surratt and Goddard [27] for the ground state out-of-plane bending of CH_{3} and CH_{3}^{+}, 585 and 1570 cm^{-1}, respectively. The vibronic structures of Rydberg states in general are expected to be closer to those the ground state ions. The values of ν_{2} for CH_{3}^{+} and CD_{3}^{+} measured by Dyke et al. [14] by VUV photon-electron spectroscopy, for example, are 1380 and 1050 cm^{-1}, respectively. These values are close to those given in Table VI.

2. Two-photon resonance enhanced ionization

The Rydberg transitions of CH_{3} and CD_{3} detected in the spectral region 260-360 nm by REMPI were identified as those allowed only by 2-photon excitation. On the basis of the results of a polarization study and of a detailed rotational analysis of the observed vibronic fine structures near 333 nm by Hudgens [39], one of the two new Rydberg series discovered in this spectral region was assigned as the np $^{2}A_{2}''$ states. The other appeared in the 265-277 nm region and was tentatively identified with the nf $^{2}E'$ states.

Figure 7 shows the 2-photon resonance enhanced MPI spectra of CH_{3} and CD_{3} in the 325-335 nm region. These spectra were obtained by using DTBP and DMSO-d_{6} as the sources of CH_{3} and CD_{3}, respectively. As is evident from these results, the intensities of the transitions occurring at 333 nm for CH_{3} and 334 nm for CD_{3} are very strong. They are about two orders of magnitude stronger than those observed in the visible region from 3-photon resonance enhanced ionization. It is also evident, particularly from the CD_3 spectrum shown in Figure 7b, that vibronic fine structure appears on both sides of the sharp peak. The expanded spectrum for this region is shown in Figure 8 for the CD_3 radical. Also shown in this figure is a computer-simulated spectrum whose construction will be described later. In order to assign the observed transitions unambiguously, further scanning over a broader spectral region was carried out. Several new peaks were detected in the shorter wavelength region 280-320 nm; they are reproduced in Figures 9 and 10 for both CH_3 and CD_3. At even shorter wavelengths (266-277 nm), additional weaker and broader transitions were observed (see Figures 11 and 12). These higher energy transitions, as will be discussed later, were attributable to resonances arising from the new nf $^2E''$ Rydberg states.

The unambiguous assignment of these new Rydberg states depended upon the resolution of the rotational structures of the bands. Comparison of the CH_3 and CD_3 spectra also proved to be a valuable tool for assignment. The observed resonances given above are listed in Table VII, along with the deuterium isotope shifts. The band origins were identified from their relatively small isotope shifts. Observed levels corresponded to Rydberg series of n = 3,4,5,6 for CD_3 and n = 3,4 for CH_3. The observed quantum defects of 0.6 are consistent with those expected for the excitation of an electron from the $2pa_2''$ orbital of the ground state to an np Rydberg orbital [39]. Further confirmation of this assignment was obtained from a polarization study and rotational analysis.

Most of the spectra obtained by REMPI for CH_3 and CD_3 showed only a strong Q branch with no resolvable rotational structure. The one notable exception was the CD_3 band at 333.9 nm (see above), for which S, R, Q, P, and

O branches were sufficiently well resolved to permit analysis. The rotational analysis performed by Hudgens [39] assuming that the 333.9 nm CD_3 band shown in Figure 8 derived from the 3p $^2A_2''$ <--- $\tilde{X}$ $^2A_2''$, 0_0^0 transition agrees very well with the observed fine structure as indicated in the figure. In this simulation calculation, the line strengths tabulated by Chen and Yeung [40] were employed and the rotational temperature of the CD_3 was taken as 300 K. The observed very intense Q-branch transitions are also consistent with the $\Delta K = 0$ selection rule for the $^2A_2''$ <--- $^2A_2''$ transition (see Table I). An additional test for the effect of laser beam polarization on ion signals was also made [39]. The observed decrease in the Q-branch intensities of CH_3 and CD_3 around 333 nm as the beam polarization was changed from linear to circular further supports the $^2A_2''$ <--- $^2A_2''$ assignment. $^2E'$ <--- $^2A_2''$ transitions are expected to show an increase rather than decrease in ion signal as the laser beam is changed from linear to circular polarization [39,40].

Most of the vibrational structure observed was attributed to the ν_2 out-of-plane bending mode. Since this frequency increases when the electron is promoted to an excited Rydberg state, higher vibronic levels (such as 2_1^1, 2_2^2, etc.) are shifted to the blue with respect to the origin of the 0_0^0 band. A sharp band at 317.9 nm in the CH_3 spectra shown in Figure 9 did not fit into the ν_2 sequence and was attributed to the ν_1 symmetric stretching mode. The possibility that this feature was due to a new electronic state was ruled out from the fact that it did not appear in the CD_3 spectrum [39]. The magnitudes of isotopic shifts estimated for this stretching (1_0^1) mode and the 2_0^2 transition of CH_3 as indicated in the figure account for the broadening of the 323 nm CD_3 peak. Table VIII shows the vibrational spacings deduced from the methyl radical spectra.

There are several possible Rydberg series that may be assigned to the transitions appearing in Figures 11 and 12 according to the selection rules given in Table I. Among these are the nd $^2E'$, np $^2E'$, and nf $^2E'$ series. On the basis of the expected values of quantum defect and additional lower members of Rydberg states expected to exist at longer wavelengths which were not detected experimentally, the observed 276 and 267 nm bands were attributed to the 4f $^2E'$ <--- $\tilde{X}$ $^2A_2''$ and 5f $^2E'$ <--- $\tilde{X}$ $^2A_2''$ Rydberg transitions with zero quantum defect [39]. The former is the first member of the nf $^2E'$ series as expected; this eliminates the difficulty of finding missing peaks associated with the nd $^2E'$ and np $^2E'$ assignments [39].

The result of band contour analyses for the 276 nm band of both CH_3 and CD_3 as shown in Figure 11, based on the 4f $^2E'$ assignment, also agrees qualitatively with the experimental profiles [39].

3. One-photon resonance enhanced ionization

The one-photon resonance enhanced ionization of CH_3 and CD_3 was first demonstrated by Welge and coworkers [41], employing a tunable dye laser operating at 215 nm for the $\tilde{B}$ $^2A_1'$ <--- $\tilde{X}$ $^2A_2''$ resonance excitation and a quadrupled Nd:YAG output at 266 nm for both ionization and initial radical generation. The 266 nm laser had energies up to 10 mJ/pulse with a bandwidth of about 8 cm^{-1} and pulse duration of 6-7 ns. The tunable 215 nm excitation laser was obtained by Raman shifting of a frequency-doubled dye laser output at 294 nm using high-pressure H_2 gas. This excitation source had energies up to 30 uJ with a pulse width of about 5 ns.

Methyl radicals were generated by the photodissociation of methyl iodide (CH$_3$I and CD$_3$I) at 266 nm using the same high energy ionization beam. The sequence of the photodissociation and ionization processes can be mechanistically represented as follows using the CD$_3$ radical as an example [41]:

Radical generation:

$$CD_3I + h\nu(266 \text{ nm}) \longrightarrow CD_3(\tilde{X}\ ^2A_2'', v'') + I(^2P_{1/2}, _{3/2})$$

Resonance excitation:

$$CD_3(\tilde{X}\ ^2A_2'', v''=0) + h\nu(215 \text{ nm}) \longrightarrow CD_3(\tilde{B}\ ^2A_1', v'=0)$$

Nonresonance photoionization:

$$CD_3(\tilde{B}\ ^2A_1', v'=0) + h\nu(266 \text{ nm}) \longrightarrow CD_3^+ + e^-$$

Since the lifetimes of the $\tilde{B}\ ^2A_1'$ Rydberg state of CH$_3$ and CD$_3$ are very short [41], the overlapping of the two laser pulses is critical. Figure 13 shows the MPI mass spectra of CD$_3$ obtained from different combinations of the dissociation/ionization (266 nm) and resonance excitation (215 nm) laser pulses. The spectra given in Figure 13(a) and (b) illustrate the effect of exciting the Q and R branches at 214.50 and 214.41 nm, respectively. The results shown by spectra (c) and (d) clearly indicate that very little CD$_3^+$ signal was generated by the 266 or 215 nm laser pulse alone.

The intensity of CH$_3^+$ and CD$_3^+$ ions as a function of excitation wavelength is shown in Figure 14. The CH$_3^+$ spectrum is seen to be more diffuse and broad, whereas that of CD$_3^+$ is sharper and better resolved. Welge and coworkers calculated the positions of various expected transitions using Herzberg's $\tilde{B}\ ^2A_1'$ state spectral data for the CD$_3$ radical. From this calculation, the extent of rotational excitation of the nascent CD$_3$ radical from the 266 nm photodissociation of CD$_3$I up to N''=10 was observed. On the basis of the spectral profile, they estimated that the CD_3 had a rotational temperature of about 250 K, carrying only 0.5 kcal/mole of rotational energy. This finding, interestingly, agrees well with the result of Lee and coworkers [42], who reported that the CD_3 radical formed by the photodissociation of CH_3I at 266 nm had only 1 kcal/mole of rotational energy. From the widths of isolated MPI spectral lines given in Figure 14, Welge and coworkers estimated the lifetimes of CH_3 and CD_3 in the $\tilde{B}$ $^2A_1'$ state to be 0.12 and 1.2 ps, respectively. The latter agrees with the value estimated by Glänzer et al. [43].

On the basis of the known radical source densities and the extent of photodissociation and product rotational excitation by the 266 nm laser, Welge and coworkers estimated the detectivity of methyl radicals by one-photon resonance enhanced ionization to be on the order of 3×10^7 particles/cc, corresponding to an absolute number of approximately 2×10^4 particles in the volume overlapped by the two laser beams. This detectivity is at least as sensitive as the LIF technique.

B. Fluoromethyl Radicals

1. CF_3 Radical

The trifluoromethyl radical is a transient species involved in such processes as etching of semiconductor surfaces [44], halocarbon fire suppression [45], and carbon-13 isotope enrichment by IR multiphoton photolysis [46-48]. It is difficult or impossible to detect CF_3 by UV absorption under most experimental conditions, and it has no known fluorescence spectrum.

Duignan, Hudgens, and Wyatt [35] used the frequency-trebled output of a Nd:YAG laser to pump the dye laser for their REMPI studies of CF_3. The dye laser beam was focused in the ionization region of the quadrupole mass spectrometer as depicted in Figure 2. Trifluoromethyl radicals were produced both by thermal decomposition of CF_3I and by IR multiphoton dissociation (IRMPD) of CF_3I, CF_3Br or CF_3COCF_3.

Figure 15 shows a comparison of results obtained by pyrolysis and by IRMPD. There is a distinctive feature in the latter spectrum at 454.3 nm that is only weakly expressed in the pyrolytic counterpart. In general, the spectra obtained from radicals generated by IRMPD are broader than the pyrolytic spectra and the signal-to-noise ratio is somewhat poorer. This is probably the result of a higher degree of internal excitation of fragments in IRMPD.

Figure 16 shows the REMPI spectrum of CF_3 in the 415-493 nm region obtained by pyrolysis of CF_3I, using four different laser dyes to cover the total region. Assuming three-photon resonance enhancement, the observed band heads are in good agreement with those reported by Basco and Hathorn [49] from their UV absorption studies. This spectrum shows a series of 13 multiplets, with an average separation of 820 $^{\pm}10$ cm^{-1} between the band heads. The four bands at 419.6, 424.5, 429.5, and 434.5 nm had not previously been attributed to CF_3. Two other bands at 433.7 and 484.1 nm, which clearly did not belong to the same procession, were also observed here for the first time. Two strong bands at 479.8 and 478.1 nm were found to be present in spectra taken with the pyrolysis oven turned off. These bands were attributed to the formation of CF_3 radicals by an alternative mechanism involving multiphoton absorption by the parent CF_3I molecule.

Interpretation of CF_3 spectra is complicated by the fact that its electronic structure is not fully understood. It is known from ESR studies [50] that the radical is pyramidal in its electronic ground state, and the adiabatic ionization potential has been obtained from the photoionization spectrum [51]. It has been proposed that the absorption spectrum in the 146-165 nm region [49] is due to a series of Rydberg transitions, possibly terminating in a 3p upper state. The origin of this transition has not been observed.

Two distinct mechanisms were considered as possible origins of the CF_3I background structure. The first involves a three-photon resonant absorption followed by photoionization and then by multiphoton dissociation of the resulting CF_3I^+ ion:

$$CF_3I + 3h\nu \longrightarrow CF_3I^* \xrightarrow{h\nu} CF_3I^+ \xrightarrow{nh\nu} CF_3^+ + I \ , \ n=1,2,\ldots.$$

According to the second mechanism, electronically excited CF_3I dissociates to form vibrationally excited CF_3, which is then photoionized:

$$CF_3I + 3h\nu \longrightarrow CF_3I^* \longrightarrow CF_3(v) + I \xrightarrow{nh\nu} CF_3^+ + e^- \ , \ n \geq 2$$

It was not possible to distinguish between these two mechanisms experimentally, although the predissociative nature of the vacuum-UV absorption bands was cited as possible evidence in support of the second mechanism.

2. CH$_2$F Radical

Very recently, Hudgens and Dulcey [52] have reported the first optical absorption of the CH$_2$F and CD$_2$F radicals by REMPI in the spectral region between 290 and 360 nm. These radicals were generated by the reaction of F atoms with CH$_2$CO and CD$_2$CO using the discharge-flow apparatus shown in Figure 3. For CH$_2$F, the F-atom abstraction reaction F + CH$_3$F ---> CH$_2$F + HF was also demonstrated to be convenient and successful (see Table II).

Strong ion signals were observed in the regions 290-310 and 340-360 nm. There was no evidence of fragmentation of the parent ions. The CH$_2$F spectrum in the 340-360 nm region is shown in Figure 17, indicating the existence of a regular progression of rotational band heads which was not observed in the CD$_2$F counterpart.

The CH$_2$F radical has an ionization potential of 9.3 ev [53]. In the range of wavelengths covered, the spectrum can be interpreted as the successive absorption of three photons, with the first two required for resonance excitation and the third for ionization. This is analogous to the mechanism proposed for the CH$_3$ band occurring in the same general spectral region. Based on this assumption, the observed band origins can be fitted to an np Rydberg series with n = 3 and 4 and a quantum defect of 0.55 [52].

C. Hydroxymethyl Radical

The hydroxymethyl radical, CH$_2$OH, is of interest because of its role in many mechanisms of combustion and air pollution. It generally is accompanied by the isomeric methoxy, CH$_3$O, which can be detected and monitored by LIF. However, CH$_2$OH has no known absorption or fluorescence spectrum which could form the basis for a detection method. Thus the demonstration of a REMPI spectrum has potentially important practical application for combustion diagnostics.

Dulcey and Hudgens [36] used a similar discharge flow/mass spectrometer system (see Figure 3) to obtain the REMPI spectrum of the hydroxymethyl radical, CH_2OH. The radicals were prepared by the reaction of F atoms with methanol. Figure 18 shows the REMPI spectrum of the mass 31 peak. The strongest signal was observed at 486.7 nm, while other significant features occurred at 468.5, 470.2, and 453.0 nm. The wavelength ranges 415-429 and 490-530 nm were also scanned, but showed no detectable REMPI signals.

No mass peaks attributable to fragmentation were observed for either CH_3OH or CD_3OD. To determine the distribution of spectral intensity between CH_2OH and CH_3O, the experiments were repeated using F + CH_3OD. In this case the two product channels give HF + CH_2OD or DF + CH_3O. From the fact that only mass 32 peaks were observed, it was concluded that hydroxymethyl was the sole source of the REMPI spectrum. Furthermore, the failure to observe a mass 31 REMPI spectrum indicated that the rate of isomerization of CH_3O to CH_2OH under the conditions of the experiment was less than 15 s^{-1}.

Hydroxymethyl radicals were produced by the reaction of F atoms with methanol. One of the complications of this method is that the F atom may extract either the hydroxy hydrogen to form methoxy radical or one of the methyl hydrogens to form hydroxymethyl. The branching fraction for these two paths has been measured by both LIF and chemiluminescent titration [54] methods, indicating that about 65% of the total product is CH_2OH.

Dulcey and Hudgens were unable to identify the resonant state of CH_2OH involved in the REMPI process because of the comparatively narrow spectral range investigated. Furthermore, it could not be determined whether it was a

one-, two-, or three-photon transition. By arguing that the most likely
candidates should be Rydberg states resulting from the excitation of an
unpaired electron from an essentially orthogonal carbon p orbital, they were
able to present a speculation based on the known ionization potential,
estimated band origin, and the Rydberg formula. On this basis, they
suggested a simultaneous two-photon transition to the 3s Rydberg state as the
source of the resonance-enhanced ionization. Verification of this tentative
assignment will require observation of other members of the same Rydberg
progression.

D. Allyl and 2-Methylallyl Radicals

Hudgens and Dulcey [55] produced allyl radicals by the reactions of F
and Cl atoms with cyclopropane and propene, obtaining REMPI spectra by
essentially the same experimental methods as before. In each case the mass
41 peak was the only one observed, and the two ion spectra were essentially
identical. The best allyl radical spectra were obtained from the reaction F
+ propene (Figure 19). These simple abstraction reactions were chosen over
other sources because there is no energy available from internal
isomerization and the radicals are cooler. The mass spectra showed no
evidence of further fragmentation of the allyl ions. REMPI spectra were also
obtained for allyl-d_6 radicals, showing that deuteration results in only
small shifts in the spectral bands, as indicated by Figure 19(b). Positive
frequency shifts were observed for two allyl peaks at 498.8 and 501.5 nm.

REMPI spectra of the 2-methylallyl radical were obtained by reacting
either F or Cl with isobutene. Identical spectra were obtained in either
case (Figure 20). The observed spectrum of 2-methylallyl radical is similar

to those of allyl and allyl-d$_6$. It is shifted by approximately 22 nm toward longer wavelengths and exhibits a more intense underlying continuum.

Significant ion currents were observed when isobutene was irradiated in the absence of the atomic reactant. The strongest peak corresponded to the molecular ion and was attributed to REMPI of isobutene itself, presumably by three-photon resonance with a previously reported transition [56]. Other mass peaks indicated that the isobutene fragmented when it ionized. However, no signal corresponding to the 2-methylallyl radical was observed in the absence of F or Cl atoms.

In the case of Cl + isobutene, the formation of 2-methylallyl is the only thermochemically allowed product channel. With F atoms, however, the formation of fluoroethylene and methyl radical or 2-methylpropenyl radical are also thermochemically accessible. The former of these alternative channels was detected by the three-photon REMPI of methyl radicals at about 450 nm.

Both the allyl and 2-methylallyl radicals are bent, belonging to the C_{2v} point group [57-61]. In each case the electronic ground state is of A_2 symmetry, corresponding to a single unpaired electron in a molecular orbital of a_2 symmetry. For a ground state of A_2 symmetry in the C_{2v} point group, single-photon transitions are allowed to states of A_2, B_1, and B_2 symmetry. Only A_1 states are symmetry forbidden. Single-photon transitions to both valence [62] and Rydberg [63] $^2B_1$ states have been observed by flash photolysis. For simultaneous absorption of two or more photons, all excited states become symmetry-allowed.

The number of photons absorbed in the preparation of the excited states was deduced by starting with the known ionization potentials of allyl and 2-methylallyl radicals. For allyl, ionization can occur after the absorption of four identical photons of energy greater than 610 nm. Single-photon resonance enhancement was ruled out because of the absence of an allowed excited state of sufficiently low energy. Since the higher excited states of both allyl and 2-methylallyl are essentially Rydberg in character, Hudgens and Dulcey were able to deduce the number of photons absorbed from a comparison of the electronic origins of the spectral progressions of the two radicals. In particular, they used the relationship

$$m = (IP_a - IP_b)/(V_a - V_b) \quad ,$$

where m is the number of photons absorbed simultaneously, IP is the ionization potential, V is the laser fundamental frequency at the electronic origin, and a and b refer to allyl and 2-methylallyl radical, respectively. The use of this relationship was justified by the argument that the shift of the band origin is due primarily to the difference in ionization potential of the two radicals.

The results of this analysis led to the conclusion that the process involved is a simultaneous two-photon absorption into a resonant state. These were assigned as 3s $^2A_1$ Rydberg states for both allyl and 2-methylallyl. As pointed out above, they are symmetry-forbidden to single-photon transitions. Ionization of these two-photon excited states thus requires the absorption of two additional photons of the same energy.

The REMPI spectra of both allyl and 2-methylallyl showed vibronic structure similar to that observed by photoelectron spectroscopy. This is consistent with the argument that the multiphoton excitation of an essentially non-bonding electron to a weakly-interacting Rydberg state has almost the same effect as removing it entirely.

IV. Concluding Remarks

In this chapter we have reviewed the technique of resonance enhanced multiphoton ionization (REMPI) and its application to the detection and spectroscopy of nonfluorescing radicals, specifically methyl and substituted methyl radicals (CH_2OH, CF_3, CH_2F, $CH_2C_2H_3$, and $CH_2C_3H_5$). We have placed primary emphasis on the methyl radical, for which 1-3 photon REMPI data are now available with some detailed rotational fine structure information. To aid the understanding of the observed REMPI spectral data, a discussion of the selection rules for 1-3 photon resonance excitation has been included.

The results of 1-photon REMPI by Welge and coworkers [41] and 3-photon REMPI by DiGiuseppe, et al. [10,11] have confirmed most of Herzberg's UV and VUV absorption data for the β, γ, and δ Rydberg series [13,32]. These Rydberg transitions are allowed by either 1- or 3-photon excitation. Two Rydberg states detected by Hudgens et al. [39] by means of 2-photon REMPI, however, are new and not allowed by either 1- or 3-photon excitation. This discovery illustrates the utility and power of REMPI as a tool not only for radical detection but also for acquisition of new spectroscopic data. Because of the great intensity of CH_3^+ ion signals observed from the $3p\ ^2A_2''$ <--- $\tilde{X}\ ^2A_2''$ Q-branch (0_0^0) Rydberg transitions by 2-photon REMPI in the 330 nm region, these transitions can be conveniently employed for CH_3 kinetic studies. For example, they can be used to elucidate the mechanisms of OMCVD (organometallic chemical vapor deposition) processes for which metal methyls such as $Ga(CH_3)_3$, $Al(CH_3)_3$, etc. are commonly used as elemental sources. In a preliminary experiment recently carried out at NRL, we have shown by 2-photon REMPI and e^--impact detection of products of $Al(CH_3)_3$ decomposition on hot substrate surfaces that CH_3 is the only gaseous product formed in the OMCVD process [64]. Recently Rockney and Grant [65] have successfully employed 3-photon REMPI of CH_3 via the $\tilde{C} \longleftarrow \tilde{X}$ transition to study the dynamics of CH_3NO_2 photodissociation by IRMPD. The extension of this method to other systems such as high temperature combustion and flames should be straightforward.

Our coverage of the substituted methyl radicals has been brief and cursory. This is due, in part, to the absence of detailed spectral data from either conventional 1-photon UV absorption or REMPI studies. Nevertheless, the results presented in this brief discussion provide further demonstrations of the usefulness of the REMPI method as a sensitive and versatile technique for obtaining hitherto unattainable spectroscopic information.

The REMPI technique has already been established as a valuable tool for studying the spectroscopy of stable molecules [4,5]. As the results reviewed in this chapter suggest, it may prove to be even more useful for the detection and spectroscopy of "dark" transient species. It is at least as sensitive as laser-induced fluorescence and it is clearly a more general and versatile diagnostic method. With the employment of a more efficient ion collection method such as the one commonly used in TOF mass spectrometers, the detection of species at concentrations down to the order of 10^3 particles/cc by REMPI is expected to be feasible.

VI. <u>References</u>

1. R. N. Zare and P. J. Dagdigian, Science <u>185</u>, 739 (1974).

2. J. L. Kinsey, Ann. Rev. Phys. Chem. <u>28</u>, 349 (1977).

3. M. C. Lin and J. R. McDonald, in "Reactive Intermediates in the Gas Phase: Generation and Monitoring," D. W. Setser, Ed., Academic Press, p. 233, 1979.

4. D. M. Friedrich and W. M. McClain, Ann. Rev. Phys. Chem. <u>31</u>, 559 (1980).

5. P. M. Johnson and C. E. Otis, Ann. Rev. Phys. Chem. <u>32</u>, 139 (1981).

6. J. W. Fleming, M. E. Umstead, and M. C. Lin, unpublished results.

7. J. T. Herron and R. E. Huie, J. Phys. Chem. Ref. Data <u>2</u>, 467 (1974).

8. (a) R. J. Buss, R. J. Baseman, G. He, and Y. T. Lee, J. Photochem. <u>17</u>, 389 (1981); (b) H. E. Hunziker, H. Kneppe, and H. R. Wendt, J. Photochem. <u>17</u>, 377 (1981); (c) K. Kleinermanns and A. C. Luntz, J. Phys. Chem. <u>85</u>, 1966 (1981).

9. R. J. Cvetanovic and D. L. Singleton, Rev. Chem. Interm. <u>5</u>, 183 (1984).

10. T. G. DiGiuseppe, J. W. Hudgens, and M. C. Lin, Chem. Phys. Lett. <u>82</u>, 267 (1981).

11. T. G. DiGiuseppe, J. W. Hudgens, and M. C. Lin, J. Phys. Chem. <u>86</u>, 36 (1982).

12. F. A. Paneth and W. Hofeditz, Chem. Ber. <u>62B</u>, 1335 (1929).

13. G. Herzberg, Proc. R. Soc. London Ser. A <u>262</u>, 291 (1961).

14. J. Dyke, N. Jonathan, E. Lee, and A. Morris, J. Chem. Soc. Faraday Trans. 2 <u>72</u>, 1385 (1976).

15. T. Loenig, T. Balle, and J. Snell, J. Am. Chem. Soc. <u>97</u>, 662 (1975).

16. G. B. Ellison, P. C. Engelking, and W. C. Lineberger, J. Am. Chem. Soc. <u>100</u>, 2557 (1978).

17. M. Karplus, J. Chem. Phys. <u>30</u>, 15 (1959).

18. R. W. Fessenden, J. Phys. Chem. <u>71</u>, 74 (1967).

19. L. Y. Tan, A. M. Winer, and G. C. Pimentel, J. Chem. Phys. <u>57</u>, 4028 (1972).

20. C. Yamada, E. Hirota, and K. Kawaguchi, J. Chem. Phys. 75, 5256 (1981).

21. H. W. Hermann and S. R. Leone, J. Chem. Phys. 76, 4759 (1982).

22. (a) D. E. Milligan and M. E. Jacox, J. Chem. Phys. 47, 5146 (1967); (b) M. E. Jacox, J. Mol. Spectrosc. 66, 272 (1977).

23. J. M. Riveros, J. Chem. Phys. 51, 1269 (1969).

24. A. Snelson, J. Phys. Chem. 74, 537 (1970).

25. J. Pacansky and J. Bargon, J. Am. Chem. Soc. 97, 6896 (1975).

26. R. McDiarmid, Theor. Chim. Acta (Berlin) 20, 382 (1971).

27. G. T. Surratt and W. A. Goddard III, Chem. Phys. 23, 39 (1977).

28. K. Morokuma, L. Pedersen, and M. Karplus, J. Chem. Phys. 48, 4801 (1968).

29. P. Millie and G. Berthier, Int. J. Quantum Chem. Symp. 2, 67, (1968).

30. Y. Ellinger, G. Pauzat, V. Barone, J. Donady, and R. Subra, J. Chem. Phys. 72, 6390 (1980).

31. J. Pacansky, J. Phys. Chem. 86, 485 (1982).

32. G. Herzberg, Molecular Spectra and Molecular Structure (Van Nostrand, New York, 1966), Vol. III.

33. I. M. Mills, Mol. Phys. 7, 549 (1964).

34. T. G. DiGiuseppe, J. W. Hudgens, and M. C. Lin, unpublished work.

35. M. T. Duignan, J. W. Hudgens, and J. W. Wyatt, J. Phys. Chem. 86, 4156 (1982).

36. C. S. Dulcey and J. W. Hudgens, J. Phys. Chem. 87, 2296 (1983).

37. G. K. Anderson and S. H. Bauer, J. Phys. Chem. 81, 1146 (1977).

38. L. B. Clark and W. T. Sinpson, J. Chem. Phys. 43, 3666 (1965).

39. J. W. Hudgens, T. G. DiGiuseppe, and M. C. Lin, J. Chem. Phys. 79, 571 (1983).

40. K. Chen and E. S. Yeung, J. Chem. Phys. 69, 43 (1978).

41. J. Danon, H. Zacharias, H. Rottke, and K. H. Welge, J. Chem. Phys. 76, 2399 (1982).

42. R. K. Sparks, K. Slobatake, L. R. Carlson, and Y. T. Lee, J. Chem. Phys. 75, 3838 (1981).

43. K. Glänzer, M. Quack, and J. Troe, 16th Symp. (Int.) on Combustion, The Combustion Institute, p. 949, 1977.

44. (a) J. I. Steinfeld, T. G. Anderson, C. Reiser, D. R. Denison, L. D. Hartsough, and J. R. Hollahan, J. Electrochem. Soc. 127, 514 (1980); (b) D. J. Ehrlich, R. M. Osgood, Jr., and T. F. Deutsch, Appl. Phys. Lett. 36, 698 (1980); (c) C. M. Melliar-Smith and C. J. Mogab in "Thin Film Processes", J. L. Vossen and W. Kern, Eds., Academic Press, New York, 1978, pp 497-556.

45. R. G. Gann in "Halogenated Fire Suppressants", R. G. Gann, Ed., American Chemical Society, Washington, DC, ACS Symp. Ser. 16, 318-40 (1975).

46. S. Bittenson and P. L. Houston, J. Chem. Phys. 67, 4819 (1977).

47. M. Gauthier, P. A. Hackett, and C. Willis, Chem. Phys. 45, 39 (1980).

48. P. A. Hackett, C. Willis, and M. Gauthier, J. Chem. Phys. 71, 2682 (1979).

49. N. Basco and F. G. M. Hathorn, Chem. Phys. Lett. 8, 291 (1971).

50. R. W. Fessenden and R. H. Schuler, J. Chem. Phys. 43, 2704 (1965).

51. C. Lifshitz and W. A. Chupka, J. Chem. Phys. 47, 3439 (1967).

52. J. W. Hudgens and C. S. Dulcey, "Detection of the CH_2F Radical by Resonance Enhanced Multiphoton Ionization Spectroscopy", in XVI Informal Conf. on Photochem., Harvard University, Cambridge, MA, 20-23 Aug 1984.

53. C. S. Dulcey and J. W. Hudgens, J. Am. Chem. Soc., In press.

54. D. J. Bogan, Private communication.

55. J. W. Hudgens and C. S. Dulcey, In press.

56. R. McDiarmid, J. Chem. Phys. 55, 2426 (1971).

57. R. W. Fessenden and R. H. Schuler, J. Phys. Chem. 39, 2147 (1968).

58. G. Levin and W. A. Goddard III, Theor. Chim. Acta 37, 253 (1975).

59. G. Levin and W. A. Goddard III, J. Am. Chem. Soc. 97, 1649 (1975).

60. A. F. Voter and W. A. Goddard III, Chem. Phys. 57, 253 (1981).

61. T. Takada and M. Dupuis, J. Am. Chem. Soc. 105, 1713 (1983).

62. C. L. Currie and D. A. Ramsey, J. Chem. Phys. 45, 488 (1966).

63. A. B. Callear and H. K. Lee, Trans. Faraday Soc. 64, 308 (1968).

64. D. W. Squire, C. S. Dulcey, and M. C. Lin, unpublished work.

65. B. H. Rockney and E. R. Grant, J. Chem. Phys. 79, 708 (1983).

TABLE I. Optical selection rules for transitions from the vibrationless X $^2A_2''$ state of CH$_3$ to singly excited states of different symmetries.[a]

Final vibronic ($\psi_e\psi_v$) state symmetry species	Number of simultaneously absorbed identical photons		
	One photon	Two photons	Three photons
$^2A_1'$	$\Delta K = 0$; $K=0$; $\Delta N = \pm1$ $K \neq 0$; $\Delta N = 0, \pm1$	f^b	$\Delta K = 0$ $K=0$; $\Delta N = \pm1, \pm3$ $K\neq0$; $\Delta N = 0, \pm1, \pm2, \pm3$
$^2A_1''$	f	$\Delta K = 0$ $K=0$; $\Delta N = \pm1$ $K\neq0$; $\Delta N = 0, \pm1, \pm2$	$\Delta K = \pm3$ $K''=0$; $\Delta N = \pm1, \pm3$ $K''\neq0$; $\Delta N = 0, \pm1, \pm2, \pm3$
$^2A_2'$	f	f	$\Delta K = 0$ $K=0$; $\Delta N = 0, \pm2$ $K\neq0$; $\Delta N = 0, \pm1, \pm2, \pm3$
$^2A_2''$	f	$\Delta K = 0$ $K=0$; $\Delta N = 0, \pm2$ $K\neq0$; $\Delta N = 0, \pm1, \pm2$	$\Delta K = \pm3$ $K''=0$; $\Delta N = 0, \pm2$ $K''\neq0$; $\Delta N = 0, \pm1, \pm2, \pm3$
E'	f	$\Delta K = \Delta l = \pm1$ $\Delta N = 0, \pm1, \pm2$	$\Delta K = 2(-\Delta l) = \pm2$ $\Delta N = 0, \pm1, \pm2, \pm3$
E''	$\Delta K = \Delta l = \pm1$ $\Delta N = 0, \pm1$	$\Delta K = 2(-\Delta l) = \pm2$ $\Delta N = 0, \pm1, \pm2$	$\Delta K = \Delta l = \pm1$ $\Delta N = 0, \pm1, \pm2, \pm3$

a. Hund's case (b) notation (after reference 39).   b. f = forbidden process.

Table II. Free Radical Sources

Radical	Source Reaction	Remark
CH_3	$CH_3N_2CH_3 \longrightarrow 2CH_3 + N_2$	$k(1100K)=1\times10^6 \ s^{-1}$
	$CH_3SOCH_3 \longrightarrow 2CH_3 + SO$	$k(1100K)=7\times10^3 \ s^{-1}$
	$[C(CH_3)_3]_2O_2 \longrightarrow 2CH_3 + 2CH_3COCH_3$	$k(1100K)=1\times10^8 \ s^{-1}$
	$F + CH_4 \longrightarrow CH_3 + HF$	$k(300K)=3\times10^{-12} \ cc/s$
	$CH_3I + h\gamma(266 \ nm) \longrightarrow CH_3 + I$	
CH_2OH	$F + CH_3OH \longrightarrow CH_2OH + HF$	$k(300K)\approx1\times10^{-11} \ cc/s$
CH_2CHCH_2	$F + CH_3CHCH_2 \longrightarrow CH_2C_2CH_3 + HF$	$k(300K)\approx1\times10^{-11} \ cc/s$
$CH_2C(CH_2)CH_3$	$F + (CH_3)_2CCH_2 \longrightarrow CH_2C_3H_5 + HF$	$k(300K)\approx1\times10^{-11} \ cc/s$
CF_3	$CF_3I + nh\gamma(10 \ \mu m) \longrightarrow CF_3 + I$	
	$CF_3I \longrightarrow CF_3 + I$	
CH_2F	$F + CH_3F \longrightarrow CH_2F + HF$	$k(300K)=3\times10^{-12} \ cc/s$
	$F + CH_2CO \longrightarrow CH_2F + CO$	

Table III. Dye laser wavelengths, dyes, and pumping wavelengths

Laser Wavelength (nm)[a]	Laser Dye[b]	Pumping Wavelength (nm)
257-274	Coumarin 500	355
269-278	Coumarin 540A	355
276-294	Rhodamine 574	355
282-288	Rhodamine 560	532
281-301	Rhodamine 590	532
301-318	Rhodamine 610	532
290-320	Kiton red 620	532
318-332	DODC	532
317-336	Cresyl violet 670	532
327-337	Oxazine 720	532
342-365	Nile blue 690	532
410-435	Stilbene 420	355
427-453	Coumarin 440	355
440-460	Coumarin 450	355

a. Laser wavelengths below 400 nm were obtained by frequency-doubling
with angle-tuned KDP crystals.

b. Methanol solution.

Table IV. Comparison of observed 3-photon resonance transitions with Herzberg's 1-photon VUV absorption data for CH_3 and CD_3.

Transition	CH_3 DHL[a]	CH_3 Herzberg[b]	CD_3 DHL[a]	CD_3 Herzberg[b]
$\gamma_i(0,0)$	450.8	450.8	451.3	451.2
$\delta_i(0,0)$	unresolved	448.9	unresolved	449.5
$\gamma_i(1,1)$	445.8		447.2	447.2
$\gamma_i(2,0)$	432.9	432.9	437.7	437.7
$\beta_i(0,0)$		432.7 Q+P		428.8 Qh
		431.5 R		428.4 R
$\gamma_i(3,1)$	428.4		433.4	
$\beta_i(1,1)$			unresolved	423.1 Qh
				422.7
$\beta_2(0,0)$	421.9	422.2 Qh	423.4	422.9 Qh
		421.6 R		422.4 R
$\beta_2(1,1)$			419.0	419.0
$\gamma_2(0,0)$	415.9	415.6	415.8	415.5
$\delta_2(0,0)$	unresolved	414.7	414.9	414.8

a. Data of DiGiuseppe, Hudgens, and Lin, reference 11. Wavelengths given here are in nm in air.

b. The wavelengths shown in these columns are three times the 1-photon VUV values of reference 13.

Table V. Spacings between vibronic bands of the $\tilde{C}\ ^2E''\ \leftarrow\ \tilde{X}\ ^2A_2''$ transitions.[a]

| Transition | Spacing, cm^{-1} | |
	CH_3	CD_3
$\gamma_l(1,1) - \gamma_l(0,0)$	746	609
$\gamma_l(2,0) - \gamma_l(1,1)$	2004	1455
$\gamma_l(2,0) - \gamma_l(0,0)$	2750	2064
$\gamma_l(3,1) - \gamma_l(2,0)$	726[b]	680[b]

a. From reference 11.

b. Based on tentative assignments of $\gamma_,(3,1)$ = 428.4 nm for CH_3 and 433.4 nm for CD_3.

Table VI. y_2' (in cm^{-1}) for the $\tilde{C}\ ^2E''$ Rydberg state of CH_3 and CD_3.[a]

$\Delta v_2'$	CH_3	CD_3
0 ---> 1	1349	1062
1 ---> 2	1401	1002
2 ---> 3	1329	1133

a. Data from reference 11.

Table VII. CH_3 and CD_3 Rydberg state transitions detected by 2-photon resonance enhanced ionization.[a]

| Transition | CH$_3$ | | CD$_3$ | | Isotopic shift, cm^{-1} |
	λ_{ex}, nm	2hν, cm^{-1}	λ_{ex}, nm	2hν, cm^{-1}	
3p $^2A_2''$ 0_0^0	333.4	59972	333.9	599886	−86
2_1^1	329.4	60700	330.7	60466	−234
2_2^2			327.8	60995	
2_0^2	319.1	62660	322.75	61955	−705
1_0^1	317.9	62886	322.9	61917	−969
2_1^3	315.6	63355	320.0	62473	−882
2_0^4	306.2	65300	321.7	63941	−1359
4p $^2A_2''$ 0_0^0	286.3	69837	286.5	69789	−48
5p $^2A_2''$ 0_0^0			271.5	73645	
6p $^2A_2''$ 0_0^0			264.7	75557	
4f $^2E'$ 0_0^0	275.75	72508	276.05	72431	−77
5f $^2E'$ 0_0^0	266.7	74961	267.0	74885	−76

a. Data from reference 39.

Table VIII. Measured vibrational spacings of the 3p $^2A_2''$ Rydberg state of CH_3 and CD_3.[a]

Vibrational spacing	Energy, cm^{-1}	
	CH_3	CD_3
$v_1'=0\rightarrow v_1'=1$	2914	2031[b]
$v_2'=0\rightarrow v_2'=1$	1334	1032
$v_2'=1\rightarrow v_2'=2$	1350	1037[b]
$v_2'=2\rightarrow v_2'=3$	1305	971[b]
$v_2'=3\rightarrow v_2'=4$	1336	1015

a. Data from reference 39.

b. These spacings are approximate due to perturbation by Fermi resonance.

Figure Captions

Figure 1:   Diagram showing the correlations between the states of the
             united atom (F) and the molecular orbitals of CH_3 in D_{3h}
             symmetry. States which are accessible by 1-photon excitation
             from the $\tilde{X}\ 2p^2A_2''$ ground state are identified by solid lines,
             whereas those which are allowed by 2-photon excitation are
             denoted by dashed lines. Rydberg states detected by REMPI are
             shown by the arrows (adapted from reference 39).

Figure 2:   Schematic diagram of the MPI/mass-spectrometry apparatus using
             a hot oven to generate methyl radicals (from reference 11).

Figure 3:   Schematic diagram of the MPI/mass-spectrometry apparatus using
             the discharge flow method to generate free radicals (from
             reference 36).

Figure 4:   Mass 15 (CH_3^+) 3-photon REMPI spectra in the region 432–453 nm,
             showing the effect of oven temperature. Methyl radical source:
             (a) Azomethane; (b) Dimethylsulfoxide. (Taken from reference
             11).

Figure 5:   Methyl radical 3-photon REMPI spectra observed between 430 and
             453 nm. (From reference 11).

Figure 6:   Methyl radical 3-photon REMPI spectra observed between 415 and
             436 nm. The CH_3 and CD_3 radicals were generated by the thermal
             decomposition of DTBP and DMSO-d_6, respectively. (From
             reference 11).

Figure 7:   Methyl radical 2-photon REMPI spectra observed between 325 and
             336 nm (from reference 39).

Figure 8: Experimental and simulated rotational spectra of the 3p $^2A_2''$ band of CD_3. The REMPI signal was generated by means of a circularly polarized laser beam. The intense Q-branch peak seen in Figure 7(b) is omitted here for clarity. (From reference 39).

Figure 9: Methyl radical 2-photon REMPI spectra observed between 315 and 325 nm using DMSO and DMSO-d_6 as the radical sources. Isotopic shifts for the 1_0^1 and 2_0^2 peaks of CH_3 shown are based on harmonic potential estimates. (Taken from reference 39).

Figure 10: Methyl radical 2-photon REMPI spectra observed between 284 and 288 nm using DMSO and DMSO-d_6 as the radical sources. These peaks were assigned to the 4p $^2A_2'' \longleftarrow \tilde{X} \, ^2A_2''$ transitions. (From reference 39).

Figure 11: Methyl radical 2-photon REMPI spectra observed between 275 and 277 nm using DMSO and DMSO-d_6 as the radical sources. The simulated spectra, (b), for CH_3 and CD_3 were based on the assumed 4f $^2E' \longleftarrow \tilde{X} \, ^2A_2''$ transition. (From reference 39).

Figure 12: The effect of oven temperature on methyl radical 2-photon REMPI spectra observed near 267 nm using DMSO and DMSO-d_6 as the radical sources. The oven temperature was 300 K when it was off and 1100 K when it was on. This transition was assigned to 5f $^2E' \longleftarrow \tilde{X} \, ^2A_2''$. (From reference 39).

Figure 13: REMPI spectra of CD_3I obtained with different combinations of laser sources: (a) 266 + 214.50 nm; (b) 266 + 214.41 nm; (c) 266 nm only; and (d) 214.41 nm only. (From reference 41).

Figure 14: Methyl radical 1-photon REMPI spectra observed between 214 and 217 nm using CH_3I and CD_3I as the radical sources (from reference 41).

Figure 15: Comparison of CF_3 REMPI spectra obtained from (a) thermal decomposition of CF_3I and (b) IRMPD. (From reference 35).

380

Figure 16: REMPI spectrum of CF_3 between 415 and 492 nm. The CF_3 radical was generated by the pyrolysis of CF_3I. (From reference 35).

Figure 17: REMPI spectrum of CH_2F observed between 335 and 362 nm. The CH_2F radical was generated by the reaction $F + CH_2CO$. (From reference 52).

Figure 18: REMPI spectrum of CH_2OH between 430 and 490 nm. The radical was produced by the reaction $F + CH_3OH$. (From reference 36).

Figure 19: REMPI spectra of $CH_2C_2H_3$ (allyl) and $CD_2C_2D_3$ (allyl-d_5) radicals observed between 490 and 520 nm. These radicals were generated by reactions of F atoms with propene and propene-d_6 as indicated. (From reference 55).

Figure 20: REMPI spectrum of the $CH_2C_3H_5$ (methylallyl) radical observed between 490 and 535 nm. The radical was produced by the reaction $F + isobutene$. (From reference 55).

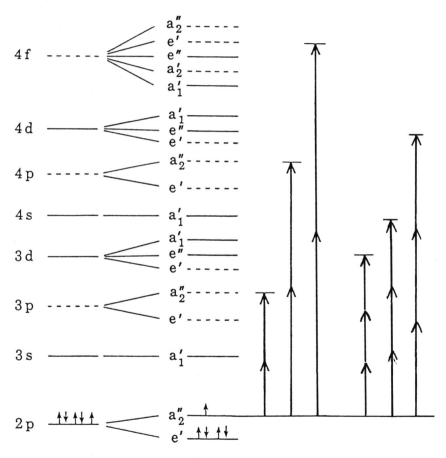

Figure 1

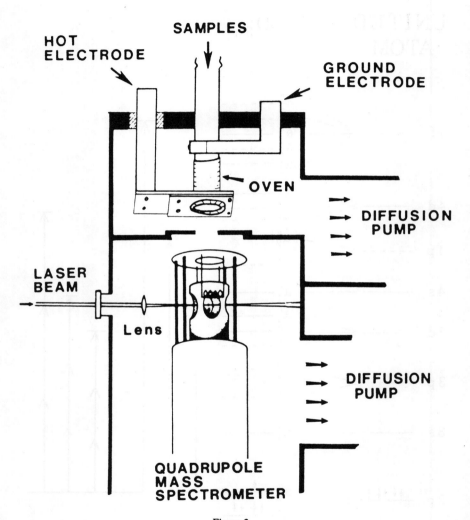

SAMPLES

HOT
ELECTRODE

GROUND
ELECTRODE

OVEN

DIFFUSION
PUMP

LASER
BEAM

Lens

DIFFUSION
PUMP

QUADRUPOLE
MASS
SPECTROMETER

Figure 2

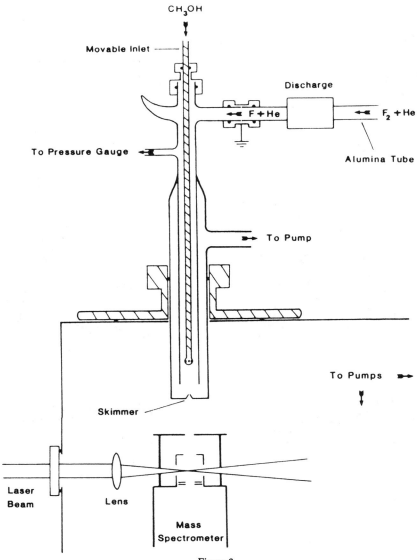

Figure 3

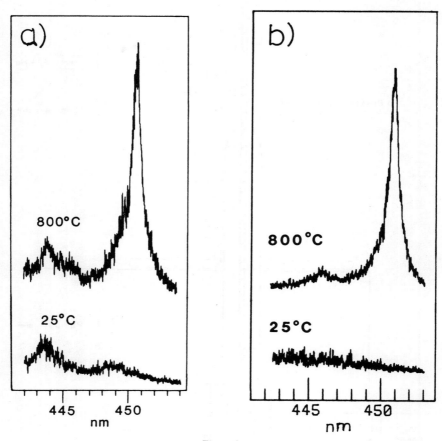

Figure 4

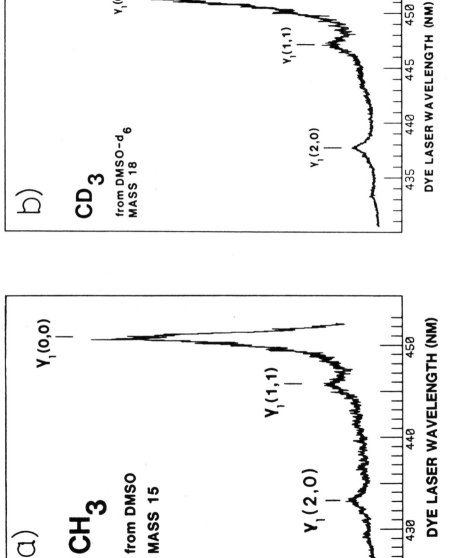

385

Figure 5

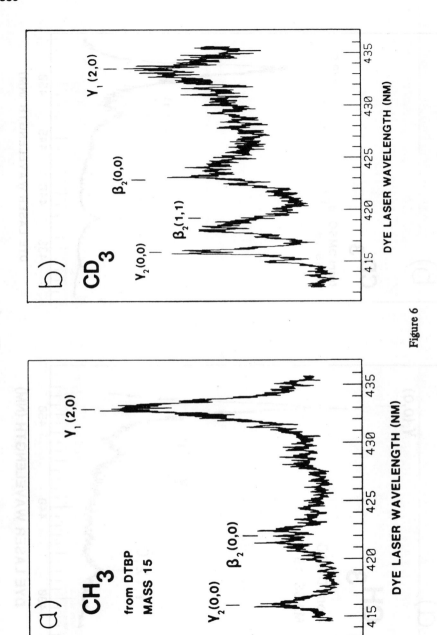

Figure 6

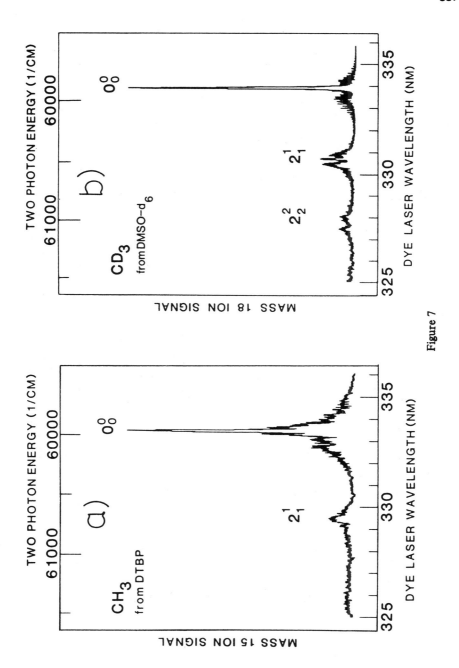

Figure 7

388

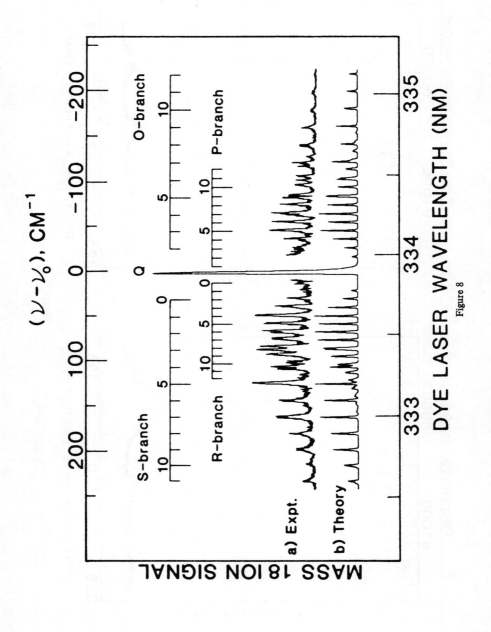

Figure 8

TWO PHOTON ENERGY (1/CM)

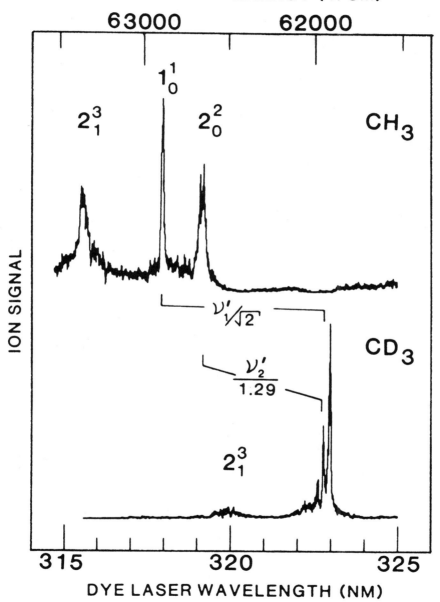

Figure 9

390

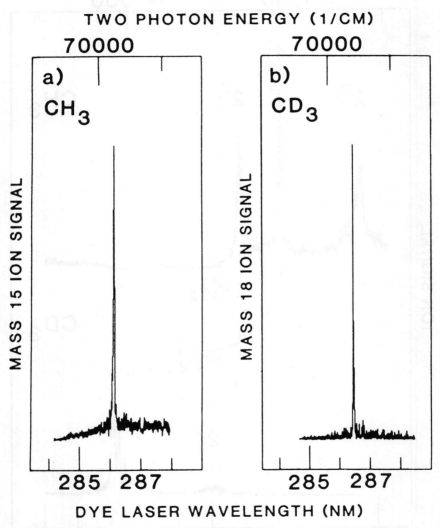

Figure 10

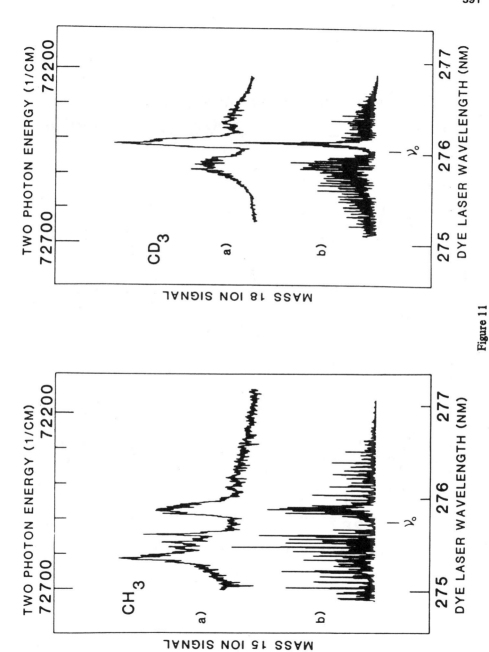

Figure 11

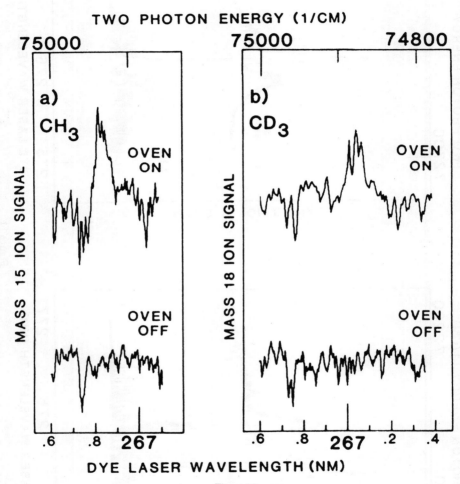

Figure 12

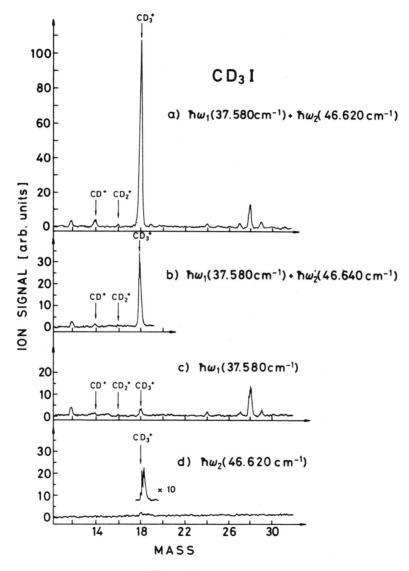

Figure 13

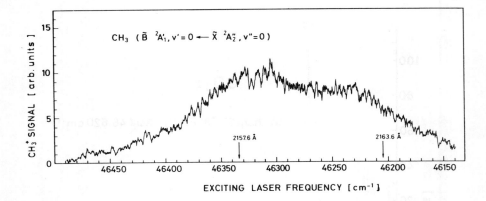

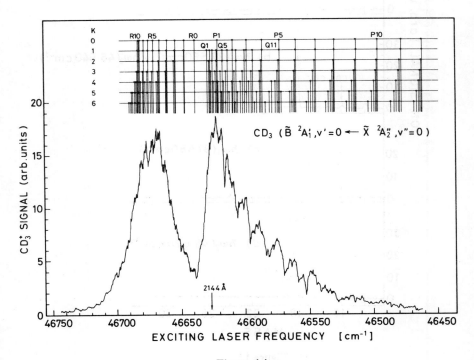

Figure 14

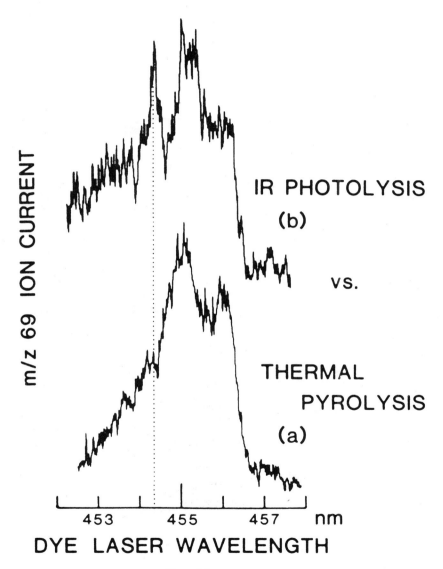

Figure 15

396

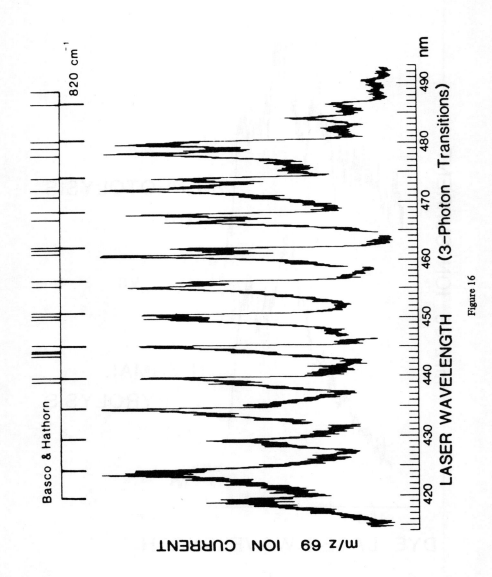

Figure 16

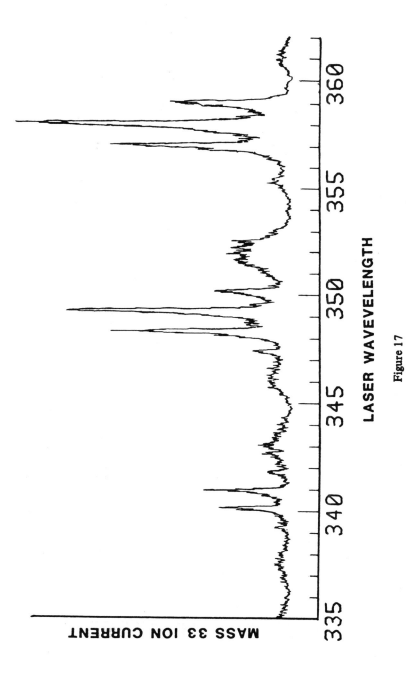

LASER WAVEVELENGTH

Figure 17

MASS 33 ION CURRENT

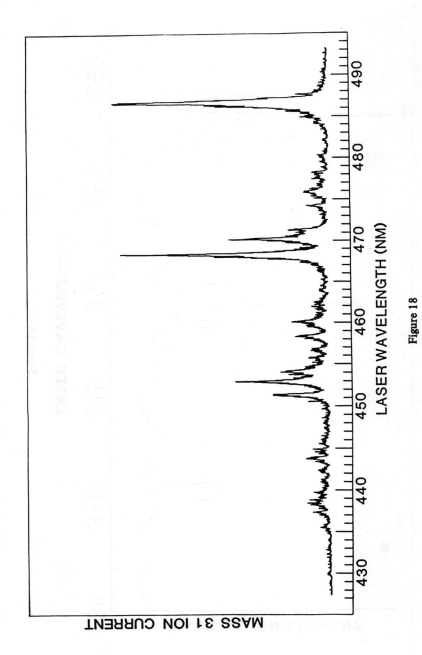

Figure 18

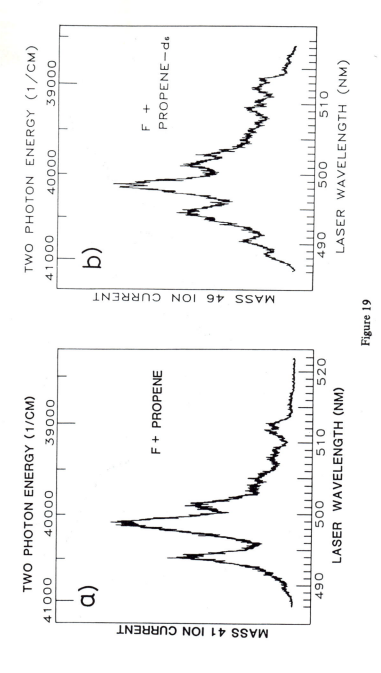

Figure 19

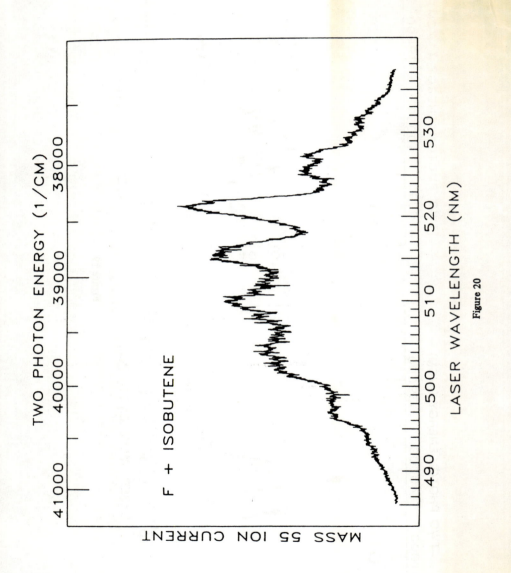

Figure 20